Shelton State Libraries
Shelton State Community College

W9-BVQ-172

DISCARDED

DISCARD

DISCARDED

# Food, Farming, and Faith

Shelton State Libraries
Shelton State Community College

SUNY series on Religion and the Environment
Harold Coward, Editor

Shelton State Libraries
Shelton State Community College

# Food, Farming, and Faith

Gary W. Fick

Shelton State Libraries
Shelton State Community College

STATE UNIVERSITY OF NEW YORK PRESS

Scripture quotations marked (ESV) are from *The Holy Bible, English Standard Version*, copyright © 2001 by Crossway Bibles, a publishing ministry of Good News Publishers. Used by permission. All rights reserved.

Scripture quotations marked (KJV) are taken from the King James Version. It was first published in 1611 and is in the public domain in the United States.

Scripture quotations marked (NASB) are taken from the *New American Standard Bible®*, copyright © 1960, 1962, 1963, 1968, 1971, 1972, 1973, 1975, 1977, and 1995 by the Lockman Foundation. Used by permission. (http://www.Lockman.org)

Scripture quotations marked (NLT) are taken from the *Holy Bible, New Living Translation®*, copyright © 1996. Used by permission of Tyndale House Publishers, Wheaton, Illinois, 60189. All rights reserved.

Scripture quotations marked (NLV) are taken from the *Holy Bible, New Life Version®*, copyright © 1969, 1976, 1978, 1983, and 1986, Christian Literature International, P.O. Box 777, Canby, OR, 97013. Used by permission.

Published by
State University of New York Press, Albany

© 2008 State University of New York

All rights reserved

Printed in the United States of America

No part of this book may be used or reproduced in any manner whatsoever without written permission. No part of this book may be stored in a retrieval system or transmitted in any form or by any means including electronic, electrostatic, magnetic tape, mechanical, photocopying, recording, or otherwise without the prior permission in writing of the publisher.

For information, contact State University of New York Press, Albany, NY
www.sunypress.edu

Production by Diane Ganeles
Marketing by Susan M. Petrie

**Library of Congress Cataloging-in-Publication Data**

Fick, Gary W.
  Food, farming, and faith / Gary W. Fick.
    p. cm. — (SUNY series on religion and the environment)
  Includes bibliographical references and index.
  ISBN 978-0-7914-7383-2 (hardcover : alk. paper)
  ISBN 978-0-7914-7384-9 (pbk. : alk. paper)
    1. Food—Religious aspects—Christianity.  2. Food habits.  3. Dinners and dining—Religious aspects—Christianity.  4. Agriculture—Religious aspects—Christianity.  5. Human ecology—Religious aspects—Christianity.  I. Title.

BR115.N87F53 2008
261.5'6—dc22                                                    2007025403

10 9 8 7 6 5 4 3 2 1

To my dad, Walter Fick Sr. (1914–2004),
the man who taught me to remember
the land and the farmer
when we thank God for our food.

Shelton State Libraries
Shelton State Community College

# Contents

# Foreword

*Food, Farming, and Faith* is a bold and welcome book that integrates ancient biblical teachings with an understanding of modern agriculture in a vital message for today's society. Gary Fick has accomplished this in a way that is easy for any reader to understand, whatever her or his relationship to agriculture and the food system it supports. It clearly is a credit to Cornell University for providing the milieu for the production of this integrative and significant interdisciplinary book.

Whether in his native state of Nebraska or my own state of Wisconsin, there are here and there across the agricultural landscape living testimonies of agricultural sustainability based upon integrated biblical and scientific principles such as Fick describes in this groundbreaking collection of twelve chapters. Among such biblically and scientifically based farmers is one I know in northern Michigan who increased his topsoil on Kalkaska Sand by an inch over twenty years and produces raw milk from his well-groomed and contented Holsteins that exceeds the standards for pasteurized milk.

Fick's message should seriously be heeded in our day as land stewardship and agricultural sustainability have largely succumbed to the pressures for immediate profits at the expense of the soil and continued fertility. Encouraged by his colleague, Richard Baer, professional agronomist Fick here presents a convincing case for "farming by the Book" while also benefiting from scientific knowledge of agriculture. On the pages that follow he relates faith to food and farming, equipping readers to take the pathway of agricultural, and cultural, sustainability.

Fick's intention in this book is to "plow deep" on connections to the Bible and faith—the part of the science and faith

spectrum that has seen so much neglect. In doing so, he aims to
have us cease eating thoughtlessly. He encourages us to exercise
our significant power by what we buy and eat. And he goes so far
as to conclude that ancient biblical farming principles point to
what is essential for agricultural sustainability in our day.

The first principle of stewardship, "The earth is the Lord's,"
brings human beings to be responsible to serve (*abad*) and protect (*shamar*) the garden, "a garden that now spreads wherever
humans go." And in this Fick envisions an agriculture that is
comprehensive and life-encompassing—an agriculture interwoven
with an understanding of wild nature and the ecological principles derived from its study.

Integratively, Fick relinks science, ethics, and praxis toward
holism—a holism that ties social, cultural, and ethical issues to the
details of nature. Powerfully, he shows that bad stewards are compelled to leave the land, even as the healing of land and life
returns. What becomes obvious is that the service the land provides us must be returned to the land with the service of our own.
This reciprocal service—a "con-service," a "con-servation"—is informed by local wisdom that pays close attention to the land as
teacher. Caring stewardship is dynamic, ever responding to the
consequences of the farmer's actions on the land, the soil, and its
life. Served by the soil as a great "ecological servant," the farmer
reciprocates by returning service to the land—a con-servation.
And in a robust conservation, Fick shares with us both from
agronomy and from the Bible, showing that the best systems will
mimic the proven designs of wild nature, meaning that we need
to be keen students of nature.

"Systems thinking" is a highly important contribution of this
book—thinking about soils, drainage, fertilizers, crops, cows,
markets, fairs, churches, schools, and more, altogether and inter-
actively, in order to see the big picture. Such holistic thinking is
vital to agricultural sustainability and to all who eat from the fruit-
fulness of the land.

The biblical picture of the farmer is not just of one who
knows the methods and the science but of one who is in a prayerful, attentive relationship with the Creator. It is a farmer that seeks
first the Kingdom of God—seeks the integrity of the farm and the
landscape in perpetuity. Agriculture, and agrarian culture, em-

braces human culture. And this has implications for farm size—farms can be no larger than what can be nurtured with love and care. Remarkably in this regard, in his extensive probing of the scriptures, Fick finds Isaiah 5:8, "Destruction is certain for you who buy up property so others have no place to live. Your homes are built on great estates so you can be alone in the land."

What is in store for you as a reader of this groundbreaking book? It is an agriculture, and a kind of shopping and buying, based upon an integrated scientific, ethical, and biblical understanding of agriculture and agricultural sustainability. This understanding aims to meet the present and future needs of people for food and health while also caring for their natural and social environment.

Calvin B. DeWitt

# Preface

In February 2002, I was asked to give a talk linking my professional work in agriculture to my personal faith as a Christian. The presentation was full of supporting Bible quotations, and a colleague and friend, Dr. Richard Baer, responded that there was a book in what I had to say. As I contemplated that comment, I realized that there was a need for a book relating food and farming to the Bible. Our whole food system is changing as the counteracting forces of globalization and expensive energy affect economies and resources, and religious ideas can provide some light that will help us get safely through these challenging changes. Here I use religion in the broad, functional sense as referring to those worldview beliefs that determine cultural values and individual motivation and behavior. As I looked over the bookshelves in our libraries, I found very little that linked religious belief to the way we ought to eat or to the way we ought to farm. A holistic perspective that encompasses the values of religion is a part of the foundation for developing and adopting more sustainable eating and farming practices. Thus there is a need to address the challenges of change in a way that includes religious belief.

The religious belief represented here is Christian, and the following chapters delve into the linkage between the Judeo-Christian foundations or principles found in the Bible and agricultural sustainability as it is being developed through the art and science of farming. In a phrase, this is "farming by the Book." But in a more general sense, this study also is a model of how any religion or worldview could be analyzed in terms of its implications for food selection and production. I leave it to persons of other faiths to insert their scriptures that correspond to the Bible quotations selected in this study. Buddhist, Hindu, and Muslim groups are certainly addressing environmental issues, as can be ascertained by a search on the World Wide Web. As noted in the introduction, secular organizations also are addressing the functional relationship between religious beliefs and the solution of environmental problems.

This book is about food, farming, and faith, integrating those topics in nontechnical language for the nonspecialist. It assumes only a good basic education. Its overall objective is to increase the effectiveness of our efforts to build a more sustainable world. The part of the book about food should help us consider the consequences of our eating habits and then to act accordingly. In that sense, this is a book for everyone, because we all eat and make dietary choices. The part of the book about farming is written primarily for agricultural and environmental leaders who are looking for a simple and practical summary of relevant agricultural knowledge. The masses of agricultural information are reduced to fifteen essentials and explained in terms simple enough for most people to begin to understand how their eating habits have impacts on agricultural sustainability. The part of the book about faith interconnects food and agriculture with a religiously based motivation to become more sustainable. Sustainability requires change, and for people to change to more sustainable patterns of living, they must be knowledgeable and motivated. Faith or worldview is a great motivational force. Christians, including Christian farmers, should find this book particularly interesting, because it will help them think about how their faith relates to food and farming. Persons of all faiths, or of no faith at all, should find here a model of how a religious belief system can inform and motivate transformation to greater sustainability. My initial assertion was that this book is for everyone because it is about food. If that claims too much, then in the end the part about food informs and equips all readers to live in a way that fosters sustainability.

Each of the twelve chapters is written to stand alone. Cross-references to other chapters are provided when the topics clearly overlap, but there also is a degree of deliberate redundancy so that the chapters need not be read in order. Except for the first and the last, each chapter is interwoven with biblical references related to its topic. The first chapter focuses on the central theme of the whole book, which is eating and food. The second explores stewardship, a concept that explains much of the biblical worldview related to the environment and farming. The third chapter addresses ecology, the holistic discipline applied in agricultural sustainability. Chapters 4–9 systematically identify the essentials of agriculture, ranging from the natural resources represented by land and climate to the social and cultural aspects of economics and ethics. Chapter 7 includes a brief discussion of animal welfare in light of the Bible. Chapter 10 focuses on the issues of abuse, poverty, and women, linking those topics to today's food systems and to the evaluation of possible future alternatives. Chapter 11 then addresses related topics of current concern: starvation, obesity, and diet.

In the first eleven chapters, a general definition of agricultural sustainability is assumed. Sustainability simply means avoiding depletion or permanent damage to resources while we use them to meet life's basic necessities. It is to eat today so we can eat tomorrow and to farm today so we can farm tomorrow. "Tomorrow" designates an unspecified amount of time, but one covering at least several human generations. However, each succeeding chapter adds a level of detail to what must be encompassed in a more precise appreciation of the concept. Chapter 12 combines the details of the preceding discussions to give a full picture of the concept of agricultural sustainability.

Most of the technical and academic aspects of each chapter are thoroughly developed in other publications that are readily available. My intention has been to "plow shallow" where the ground has already been well worked. However, regarding connections to the Bible and to matters of faith, which are not well covered in previous work, my intention has been to "plow deep" and provide a thorough coverage of the ground. The result is that the number of references to books and scientific papers cited in the Notes section has been deliberately minimized. References to Bible stories and passages have been emphasized, often included as part of the text. Most of the quotations from the Bible were taken from the two most recent translations into English, the 2001 English Standard Version (ESV) and the 1996 New Living Translation (NLT). The ESV emphasizes literal accuracy. The NLT sometimes better captures the meaning of poetry and symbolism.

The Bible has been used both to rationalize and indict practices that are abusive of natural resources and people. However, both the excusers and accusers often have held too narrow a perspective on the Bible. I have tried to be comprehensive in my selection of scriptures and careful in my interpretations. Regarding my interpretations, I want to be clear, cautious, and humble. I certainly recognize that alternative and scholarly correct interpretations are possible. At the same time, my hope is that my broad compilation of relevant passages and suggested interpretations will show both (1) that some selective and limited scriptural perspectives are inconsistent and incorrect, and (2) that there is more to learn about what the Bible has to say regarding food, farming, and faith. For those concerned with biblical interpretation, my hope is that they will be encouraged to study further. Some readers may wish to verify for themselves Bible quotations. Two short appendices describing the organization of the Bible are included for novices, and an index of quoted Bible passages will be helpful for the more experienced Bible student.

I am grateful for each person who helped make this work possible. Christian author and leader Dennis Peacocke first prompted me to fully develop my ideas about the relationship of my work and faith. A few years later, Richard Baer at Cornell University encouraged me to record those ideas in writing. Calvin DeWitt at the University of Wisconsin and all of the agricultural faculty of Dordt College in Iowa listened to my initial plans and urged me to press on. The College of Agriculture and Life Sciences at Cornell University granted me a six-month sabbatical leave during which time the first draft of this book was written. David Andrews of the National Catholic Rural Life Conference and my pastor, Raymond Crognale, provided important support along the way. Encouragement and useful comments also came from Robb De Haan, chair of environmental studies at Dordt College, John Doran, former president of the Soil Science Society of America, and Peter McDonald, a friend who is a farmer and conference speaker. My wife, Mae Ellen, and colleague, David Bwamiki, read and commented on the original drafts of each chapter. Friends and associates Julie Dawson and Hugh Gauch Jr. helped by making useful editorial suggestions on the whole manuscript, and Linda Marco and Leslie Stow double-checked the accuracy of the scriptural citations (though any errors are my responsibility). Julie Dawson, Julie Grossman, and Pam Bateman kindly provided detailed advice from the female perspective that helped improve Chapter 10. Joe Regenstein of Cornell University made constructive comments from a Jewish perspective. Finally, a group of eleven students read and discussed a draft of the text as part of a special topics course I taught at Cornell. They are Victoria Christian, Danielle Farmer, Charles Fick, Collin Haight, Alphina Jui-Jung Ho, Karry Kayu Lai, Peng Li, Elizabeth Pullen, Michelle Raczka, Ian Toevs, and Michelle Dan Wang. The text is both clearer and more complete because of their contributions. I extend a special thank you to each person who has helped make this a better book. My wife, Mae Ellen, deserves special gratitude for her patience and cheer during this project that took much more time than we expected.

# Introduction

This is a little book, but it has a big purpose: that we would cease to eat thoughtlessly. We all eat, so there is value for everyone in understanding what is written here. And this is the reason: Every day when we get our food we are influencing the kind of food system that will provide our food tomorrow. Will it be healthful? Will there be enough? Will others be better or worse off by the way we eat? When we select our food, we are influencing farming and the environmental impact and sustainability of farming. We have significant power through the way we eat, and we ought to think about it.

In my college days, protestors chanted "Power to the people." Now my chant is "Power to the eaters." Interestingly, there is a complete overlap between "the people" and "the eaters." We all need to eat, but we do not all know or exercise our eating power in a knowledgeable way. Eating power is the influence that our food selection habits have on the future direction of our food system—of all those interrelated factors that work together to produce, prepare, and deliver what we eat. In a free economy, we vote for the kind of world we want by the way we spend our money. Money spent on food sends a signal that accelerates or redirects the changes that are going on all of the time in the food system. This principle also applies to those who grow, process, and deliver food. Farmers and processors are spending resources and sending economic signals through the food system. They are responding to signals that come from consumers. We influence the future by the way we vote with our food dollars. May we all vote wisely. Power to the eaters!

Of course it is not easy to vote wisely. To do so means that we must think about the consequences of how we eat. Because we do not want to vote for food systems that endanger tomorrow's food supply, we must especially understand what is essential for growing food. We should eat today so that we can eat tomorrow. In other words, we should vote for a sustainable food system. Beyond the essentials of food production are all of the social and moral associations of the food system affected by our vote. To vote wisely also means that we must evaluate the ecological, social, and moral aspects of eating. It means that we must be motivated to act wisely, and our motivations are linked to ethical and religious worldviews. I know of no one who said it better than self-acclaimed atheist George Bernard Shaw:

> Religion is a great force—the only real motive force in the world. But you must get a man through his own religion, not yours.[1]

As an example of the social and moral implications of eating, I began with the claim that we all have eating power. To refine this a little, my current understanding is that the power to foster change lies in the process of choice. The worldwide poverty and hunger problem (see Chapter 11) forces over 2 billion people into malnutrition, but choices have been and are being made by others that have social and moral effects on those with little opportunity to choose. Closer to home, there are those among us who must search dumpsters and garbage piles for food. What eating power can they apply? Still, their choice to look in those places reflects the probability that they will find food there. The homeless and broken have limited opportunity to earn or grow their own food. Their food choice might be having an effect, especially if it influences what others do about their situation. What would happen if there was not enough food in the garbage and they had to seek other sources? Would society ignore them and let them starve? What do we really believe about a person's right to food and how he or she should get it? What is wrong with this picture, and how should it be changed?

There are indeed consequences to our eating habits about which we should think. Our eating choices affect not only our personal health and appearance but also the health of the environment and of communities and individuals both near and far away. Here is a short list of some big food-related questions: Who should be growing our food, and where should it be grown? Is there any difference in the food quality and environmental effects of family farming and industrialized agriculture? Do pesticides and genetically modified organisms only enhance the food supply, or do some practices harm the environment and poison the future? These are difficult questions, and we have a vote in deciding how they will be answered. What do we need to know before we can vote wisely?

Several books address the social, ethical, and political aspects of food and eating, including those of Beckmann and Simon,[2] Schlosser,[3] Nestle,[4] and Jung.[5] The first, *Grace at the Table,* and the last, *Sharing Food: Christian Practices for Enjoyment,* assume religious perspectives.

The personal aspect of eating, especially dieting, is big business in the modern industrialized world. Religious motivation for dieting is explored in recent books by Webb[6] and by the Hallidays,[7] and responsible eating is advocated in the *More-with-Less Cookbook,*[8] *Simply in Season,*[9] and *Just Eating?*[10] Michael Pollan's book, *The Omnivore's Dilemma,*[11] is a hybrid account, linking the ethics of eating to personal dietary options in America today. Numerous works provide technical details about farming and farming systems, but such books generally ignore the related religious and ethical implications for how agriculture is practiced. *The Holy Earth,* written by Liberty Hyde Bailey at Cornell University in 1915, is a notable exception.[12] At mid-century, Alastair MacKay wrote *Farming and Gardening in the Bible.*[13] More recently, Evans, Vos, and Wright[14] and Graham[15] have examined the relationship of faith and agriculture. The closely related issues of environmental care are the subjects of many books, including current Christian perspectives such as *Caring for Creation,*[16] *God's Stewards,*[17] and *Sustainability and Spirituality.*[18] Many faiths are represented in the recent books by Hope and Young (*Voices of Hope in the Struggle to Save the Planet*[19]), Palmer and Finlay (*Faith in Conservation*[20]), and Gardner (*Inspiring Progress: Religious Contributions to Sustainable Development*[21]).

This overview of the literature makes it clear that there are numerous viewpoints and plenty of information about food, farming, and faith. However, no one has attempted to bring all three subjects together to identify essential knowledge in a way that is accessible to the typical eater. But such a work is needed if eaters are to be empowered as change agents of the food system. At least two things are required to empower them: They need knowledge, not just information, and they need motivation.

In the chapters that follow, my goal has been to uncover the essential principles that foster a sustainable future. This includes the definition of important checkpoints for sustainability, especially as they are related to food and farming. The knowledge discussed concerns the environment, agriculture, and society—all factors interconnected to food and our eating habits. Those are well-worked fields, but what is unique in this writing, especially for the parts on agriculture, is that they are deliberately integrated with a Christian perspective that is tied to the moral and ethical motivation of many people. This will hopefully encourage Christians to practice what they are supposed to believe. May it also be a model that will enable persons of other faiths and other worldviews to link their own scriptures and their own motivations to the patterns presented here.

On a personal note, my life began as the son of a Nebraska rancher and environmental conservationist. My career has been as a professor, a teacher, and a research scientist. As an agronomist, I have specialized in crop production and soil management, especially with the perennial forage crops used for livestock feed. I also have been an ardent student of Scripture, gaining insight from writers and teachers who focused on the Bible. In addition, I studied some of the farming traditions of First American cultures, and I have read the Qur'an, focusing on how it addresses environmental and agricultural concerns. With this background and perspective, it seems to me that a central and neglected issue in having food for the future has to do with the knowledge and motivation to sustain agriculture. At least it is on that issue that I am most qualified to contribute to the discussion.

Several reviewers of early drafts of this work found the aforementioned introduction to my background inadequate. They asked me to help them "see what I see." Thus each chapter now includes a short autobiographical account that is intended to reveal both my own approach to each topic as well as the biases that limit my perspective.

I have tried to limit detail to what I regard as necessary for an accessible and integrated account of food, farming, and faith. To do this, I have first placed agricultural sustainability in the dual context of the natural environment and human culture, both of which determine and direct the processes of agriculture. Regardless of our customs, we must eat, but the way we eat and the way we think about eating is a cultural phenomenon. It is a part of the culture of agri-culture. At the same time, biological and physical conditions limit what is possible for agriculture to produce. We need to be aware of both culture and nature if we are to understand the basic parts of our food systems, and the opening and closing chapters that follow address those two points of context.

Within this context of culture and nature are the actual processes and practices of farming. The details can be overwhelming, and much of the information about agriculture today is understandable only by specialists. The challenge of converting information to knowledge lies in discerning what is essential, so I have developed a list of the essentials of agriculture as a core contribution of this writing. What does it take for agriculture to continue to be practiced? Without being too complicated, I wanted to include enough detail so that informed eaters can vote wisely with their food choices. In other words, I wanted to give enough detail about what is essential so that the consequences of a food choice can be examined in light of the basic issues that every alternative also must address. In the following chapters, those in the middle identify the fifteen essentials of agriculture.

Searching for the essentials of agriculture turned out to be a challenging endeavor. As an agricultural professional, I tried several approaches but eventually settled on a combination of my own training and experience plus the knowledge of agriculture that can be discovered in the ancient texts of the Bible. I reasoned that if ancient farming principles are still used today, then they are probably essential.

Ancient wisdom and modern knowledge do not have perfect correspondence, but they proved to be close enough with regard to farming and agricultural sustainability for me to offer the list of essentials as a checkpoint for evaluating the impact of our eating habits on food production. Without telling the reader what to do, the list suggests a set of questions to ask about the effects of the way we eat. The religious element further offers a set of moral guidelines that can motivate and direct Christians to authentic Christian responses.

Recognizing needed change, and even agreeing that we personally need to change, is not enough if there is no motivation to change. Motivation to change is grounded in one's religion or moral beliefs, in a word, one's worldview. The inclusion of the religious element in these chapters is thus a very important and logical aspect of informing the process of change as it applies to the food system.

Another way of looking at this is to acknowledge that agricultural sustainability is inherently holistic. It includes all of the essentials of agriculture. It includes the past, the present, and the future. It includes all of the social, cultural, and ethical linkages in the food system as well. Thus religion is a natural component of holistic, sustainable thinking, and academics like me have erred when we have left it out of the picture. What we need to do instead is to "respect and connect" at the level where persons are motivated to change. So another ingredient in this writing is a sympathetic religious model of thinking about food and farming. The model used here is a Christian and biblical model, but the more general point is that religious ideas and motivations are a proper part of thinking about these important matters.

Probably the best outcome from this work would be an ongoing discussion that focuses on the future of the way we get the food we eat. It should be an open discussion that includes matters of faith as well as facts of farming, a discussion that is full of respect as well as an earnest desire to secure better tomorrows for future generations. When we eat thoughtfully in the light of knowledge and motivation, we can help build a sustainable world.

Shelton State Libraries
Shelton State Community College

# Chapter One

# It Is All about Food

It is a sad but true saying that our biggest problems are ignorance and apathy. It is certainly true of agriculture. With the bulk of the American population two or more generations removed from farming, few of us know very much about the sources of our food. In rich nations at the start of the twenty-first century, science and technology have made our food supply so abundant and inexpensive that the typical attitude about agriculture is complacency. At the same time, there are warning signs in our air, water, soil, climate, and even food itself that indicate agriculture must change— and that change is so significant that everyone will be involved.

A wise professor once told me that the secret to successful problem solving was to ask the right question. As an applied ecologist focusing on agricultural sustainability, my question is this: How can we draw attention to that essential body of knowledge and wisdom about agriculture that most of us so complacently ignore? A second question quickly follows: Once ignorance is reduced, how can we motivate people to change? To help answer these questions, I have been looking for a central theme that, like a cut diamond, would have many facets that unify and enliven all of the dimensions of life related to our basic dependence on farming. I think that theme is food. Though food may seem mundane and even self-indulgent, several times a day our body tells us otherwise. Civilization, with its science and technology, has made it possible for rich humans to live in a rather thoughtless

7

manner, but reflective and watchful persons also remind us that we live thoughtlessly at our own peril. The goal of this chapter is to open up our thinking about food, to show how it is life's central and comprehensive theme. The facets of food reflect on all of the aspects of life, and in that sense, it *is* all about food.

## Thanksgiving

The comprehensive associations of food can be illustrated by a couple of pictures from my own life. Starting with the elaborate, consider a holiday meal at our home, say a Thanksgiving meal in November. There will be a dozen or so people gathered around our table for a traditional feast. The people will be our own family plus an assortment of international students and friends unable to be with their own family on that special day. The preparation started back in May when I harvested and froze rhubarb that will now be a key ingredient for a favorite family dessert. In July, homegrown blueberries were picked and frozen for the same purpose. In July or August, we ordered a locally raised turkey available at the Ithaca Farmers' Market. A week before the celebration, we discussed together what each household would bring so that the meal would be a community affair. The evening before, I will have made a special cranberry salad with a recipe given to me by my mother. Typically, my wife and children will be busy at the same time making rhubarb and blueberry pies. Someone will see that there also is the usual pumpkin pie. On the morning of the day, my wife will prepare the turkey and place it in the oven for the several hours of cooking while other preparations take place. The kitchen crew will process the last of the season's fresh produce from our garden. Usually there will be "Long Keeper" tomatoes harvested in September but carefully stored for this day in November. If the local deer have not broken into our garden, then there might still be Swiss chard, Brussels sprouts, carrots, and kohlrabi or rutabaga. As my children have said, these are "real vegetables" and not the common fare from the supermarket. We will go to the grocer for green beans or corn or peas and for potatoes to be specially seasoned with herbs and spices. There also will be sweet potatoes, special at our home because they remind us of the years when sweet potatoes were one of the few foods our severely allergic child could eat. Our guests will be bringing some of their traditional dishes from Uganda or India or Korea. A close friend usually brings an amazing array of locally baked breads. There also will be special drinks and juices, often apple cider in season that time of year.

I am sure I have forgotten some items, but you get the picture. It is an elaborate feast, a celebration of abundance and friendship, and it is more than that too. At our house, we often will begin the meal with a word of thanksgiving from each person around the table. Then we will hold hands and sing a song, perhaps, "Creator God, Lord of the universe, thank you for creation, thank you for food." Sometimes people at our table will not share our faith but not complain in front of all that food. They at least observe and hopefully enjoy the ritual of a special occasion where they are guests. Often after the first session of feasting, there will be prayers of gratefulness, and I will read the story of the first American Thanksgiving to remind us of the history that is being commemorated. We will then retire to various social events that include clearing the table, doing dishes, and appropriate games for the age range of those who are present. We may gather around the piano as my wife or daughter-in-law plays and sing a few favorite songs. And then a few hours later, when we are able, we will gather at the table again for dessert, for those pies I mentioned earlier, usually accompanied by some ice cream. And so goes the feast.

This holiday is observed with countless variations in countless American and Canadian households. Financial offerings have been given and volunteers enlisted to repeat the theme at soup kitchens and homeless shelters so that even the poor and broken can participate in the feast. Turkey, a uniquely American poultry, is the traditional focal point, but as a child my family roasted a chicken. (I am not certain why, but perhaps that is all we could afford.) Because of food allergies, we also have had leg-of-lamb instead of turkey. There are special recipes for special celebrations around a vegetarian table. The details are not so important, but the elaborate and careful process speaks volumes about the associations of food. And that is the point: It is all about food. Food is a comprehensive theme.

On even a larger scale, this special meal shows connections to all of life's dimensions and meaning. There are aspects related to the physical and biological needs of life. Though extravagant, this meal does provide food nutrients we need to live. It is a step in the ecological cycles and flows supplying carbohydrates, proteins, fats, minerals, and vitamins that meet our natural needs while we are alive. You do not have to look too far past the table to come to the foundations of the food system with its interconnections to agriculture and the world of nature and nature's laws. Perhaps it is uncommon in America today, but for my house and for much of the world, those steps are no farther than the vegetable garden in the backyard. Although they are steps often not contemplated in our food-rich society, it is direct and obvious to go from the food on the table to the growing food in the field embedded in the milieu of natural ecosystems with their biophysical laws that cannot be broken. (See the following definitions [boxed text] of some technical terms used in these chapters.)

## Other Food Associations

Another food association is obvious in these celebratory feasts, and that is the social dimension. With family and friends gathered together in thanksgiving, eating is clearly a social act and statement. We eat our feasts with a ritual that has history and tradition. Messages of culture exist here. We eat with our friends and not usually with our enemies. Messages of diplomacy and reconciliation are here. We share our food with those we value and appreciate. Messages of social structure and caring are prevalent here. We eat in celebration as best we can afford. Economic messages are here too. Though matters of personal preference and health also affect our diet, it appears that all matters of culture and society are expressed in some way at some time in our food and in our eating.

The spiritual dimension is another food association that is clear in a celebration such as Thanksgiving. The American Thanksgiving holiday (holy day) has its roots in a religious observance. Many, perhaps most, of the traditional family-oriented practices

## Key Terms Used in This Book

**ag´ri·cul´ture.** The human endeavor to produce food and other useful products by the managed care of plants and animals; the art and practice of farming.

**e´co·sys´tem.** A distinct natural community characterized by dynamic interactions among its physical and biological components. Examples of ecosystems are a tundra, hardwood forest, savannah, seashore, or an agricultural field. The last example also is called an **ag´ro·e´co·sys´tem.**

**met´a·phor.** A figurative use of language for the purpose of providing insight by noting the similarity or likeness of objects and ideas literally different (e.g., agriculture is like a three-legged stool).

**mod´el.** A representation of some part of reality that contains enough detail to allow the study of certain aspects of what it represents.

**sus·tain´a·bil´i·ty.** That quality of a system that gives it persistence; the capacity to continue into the future; in practice, it means avoiding depletion or permanent damage to the resources of a system while they are used to meet life's basic necessities.

**sys´tem.** A part of reality or a set of concepts in which the relationships of the component parts are significant or meaningful. The parts of a system are interdependent or interactive.

**world´view.** The beliefs and values that are expressed in how one thinks and acts, including motivation for deliberate behavior, sometimes called **functional religion.**

described for our home have deliberate religious overtones. Taking the time to remember that there are things to be thankful for is an expression of a value system, or worldview, or religion, with primary or ultimate concerns and commitments. Some people may not think of these associations as spiritual or religious, but everyone is at some stage of developing and expressing moral and ethical ideals that emerge from their view of the world. What we eat and how we eat is full of meaning about what we believe and what we value.

All of these associations of food are rather easy to discern in a feast of celebration, but what about a simpler example also drawn from my own experiences? What about the late-night snack? Perhaps I have been watching television or reading a book. I will take a break and make some popcorn. I may eat it alone without a prayer or a thought of thankfulness, engrossed in whatever has kept me up so long that my body is calling for more food. Has the different context changed any of the associations of food with all of life? I think not. They are only less obvious.

The relationship of the bowl of popped corn to the crop in the field is still there with all of the ecological principles and consequences that operate in agricultural practices. The social aspects are not expressed in an immediate interaction with people, but the fact that the snack is popcorn instead of sugarcane tells a great deal about the cultural setting. The kind of snack also is related to nutritional knowledge and concerns about health that depend on the educational system of the culture. The kind of popcorn chosen is related to economic factors and to a particular setting in the processing, transportation, and marketing components of our food system.

The spiritual associations with food are almost invisible in this case, but upon reflection, one observes that the kernels of grain that were alive are dead by the time they are consumed. Each act of eating involves this principle of life. Life comes only from life, only by taking life. Even in this mundane vegetarian example, the kernels of grain have been killed. This is a large truth in most religions of the world, and it is demonstrated whenever we eat. We may never think about it, but it is true.

Whether an elaborate celebration or a simple snack, a little reflection shows that the biological, social, and spiritual associations of food are always present. All aspects of life are interconnected with food. That is why I say, "It is all about food." But it also is true that food, or rather the absence of food, is associated with death. If we are rich and affluent, then we may have the luxury of not thinking about food, but for many, it is a matter of life and death.

## A Matter of Life and Death

Early in the twenty-first century, a group of the world's leading economists came together to identify the most critical problems of humanity. Their findings are known as the Copenhagen Consensus of 2004.[1] The top five problems on the list are (1) control of HIV/AIDS, (2) providing micronutrients for the malnourished, (3) trade liberalization, (4) control of malaria, and (5) development of new agricultural technologies. Two of the top problems are clearly about food, and all five are interrelated.

Though it is essential, it is not sufficient to meet only the energy and protein needs in the human diet. Counting vitamins, minerals, trace elements, amino acids, and essential fatty acids, there are at least fifty nutrients that humans need.[2] In 2006, about 900 million people were hungry, that is, they had less energy and protein in their diet than they needed. Approximately 2 billion people on the planet were afflicted with malnutrition, lacking one or more of the other essential nutrients. The Food and Agricultural Organization of the United Nations (FAO) estimated that about 25,000 people died each day as the direct or indirect result of hunger and malnutrition. Two out of three were children.[3] Malnutrition (problem 2) and lack of food (problem 5) are certainly big problems. Even the diseases (problems 1 and 4) are made worse by shortages of food.

The suffering of hunger and malnutrition is doubly tragic because there is enough food produced in the world to feed everyone.[4] A big part of the problem is that the food is not produced where some of the people who need it live. Markets, transportation, and even moral determination must be enhanced to alleviate the tragedy of hunger. Thus trade liberalization (problem 3) and the associated development of the capacity of poor and hungry people to participate in trade can also be linked to the life-and-death issues that surround food.

Although it has frequently been overlooked, human health and well-being are essential aspects of sustainability. If a food system cannot keep people healthy and well nourished, then how can it be sustainable? Health is bound to agriculture by the human food chain. Agricultural sustainability must encompass human nutrition, a point that is being strongly addressed by my colleague, Ross Welch.[5]

If our analysis is theoretical and leisurely or if it is practical and urgent, then we still come to the same conclusion. It *is* all about food. This is a point well remembered in the following chapters. The material about stewardship, ecology, the fundamentals of agriculture, and the related issues of abuse and justice is really about food and thus is of vital concern to everyone. And because we all consume food, we all have the opportunity to positively influence the sustainability of our food systems.

# Chapter Two

# The Foundation of Stewardship

## Motivation

As I have promoted agricultural stewardship for almost forty years, I have sometimes wondered, how can we motivate people to change? For starters, what motivates me to change? When I was younger, I was motivated by public recognition. I have a small gold pin in a box of keepsakes in the attic. It is enameled with a green tree, a four-leaf clover, and the word "forestry." At the time I received it, it represented the most significant achievement of my young life. I was twelve years old and it was awarded by the county 4-H youth program for planting trees. I lived then in southern Holt County, Nebraska, an area of meadows and sandhill rangeland that probably did not support a single tree when the first European settlers arrived less than eighty years earlier. Trees were sparse in most of Nebraska, but Nebraska became known as "the tree-planters' state." J. Sterling Morton, who had a home in Nebraska City, initiated Arbor Day, a special day for planting trees. Tree planting was so important that the 4-H clubs of the state recognized it with an award. My motivation was rewarded with both the county and eventually the state prize for forestry. As I recall, the state award paid my expenses for a week to explore career opportunities at the University of Nebraska.

Though a prize was part of my motivation, it probably was not the most important part. There also was my parents' example.

The survival and well-being of their small herd of beef cows depended in part on shelter from winter storms. An adequate, most affordable shelter was a good grove of trees. Some of my earliest memories are of following them as my mother carried baby trees with roots submerged in a bucket of water as my father planted them one by one in the sandy soil. In a few years, I was carrying the bucket. In a few more years, I was planting trees beside my father as my little brother carried the bucket. Eventually I planted a grove of about three acres of trees with my dad carrying the bucket. That became my prize-winning "shelter belt." Today, those trees are grown and protect my brother's cows from bitter winter winds.

Weeds were a big obstacle in the process of getting the trees to survive and grow. We would go out each summer and hoe the weeds around the trees. A rainy spell in summer meant a break from making hay in order to hoe trees. My father wore out a garden hoe by chopping down weeds around those trees. That "worn-out hoe" with a paper-thin blade is now on display in my office with an accompanying story. Dad continued to plant and hoe trees all of his life. A few weeks before he died at ninety years of age, he was out planting trees. He told me, "You do not plant trees for yourself. You plant trees for the next generation, for the future." My parents' example and wisdom certainly motivated me.

But, in general, what motivates people? As I started to work with organic farmers and others looking for a more sustainable way of doing agriculture, and this was years ago when it usually involved extra personal and financial sacrifices, I observed that almost all of these people were motivated by an ethic of stewardship, by a worldview that prompted them to live so as to take care of the earth and other people as best they could. They did not always answer my question about motivation in religious terms, but sometimes they did. A significant number of these farmers were Christians who wanted their farming methods to represent their faith. My father did not practice "church religion" and did not give an overtly religious answer when asked about his motivation. Nevertheless, I could see that he too was motivated by a functional religion that valued land and served future generations. In my academic circles, I had sometimes heard religion blamed for environmental exploitation, so it was a surprising insight for me that there was a positive connection between environmental care, food, farming, and faith. This has prompted me to examine various faith systems in terms of what they have to say about food, farming, and the environment. I have appreciated religious or worldview discussions with First Americans and international students of many faiths about the connection of agriculture and their beliefs. I also have read the Qur'an with a focus on these topics. But the main findings, what I know best and what I will emphasize here, come from my study of the Judeo-Christian tradition, the ancient wisdom about stewardship recorded in the Bible.[1]

Any consideration of the relationship of the Bible to the art and science of agriculture must include an interpretation of the story of the Creation and its initial assumptions about humans and the responsibilities they were given. The story in Genesis, the first book of the Bible, is traditionally regarded as being compiled by Moses. The assessments of the story range on a continuum from mere ancient myth to literal truth. However it is evaluated, the biblical story of the Creation represents the ancient understanding that is "truer than fact"[2] about human relationships to deity, to one another, and to the rest of creation. The Judeo-Christian worldview summarizes those relationships with the concept of stewardship. The stewardship concept is the foundation of everything else about a biblical view of agriculture, and the goal of this chapter is to offer a brief outline of that understanding.

## In the Beginning

In the first chapter of Genesis, the Creator is depicted as valuing the Creation by repeatedly declaring that it was "good" (verses 4, 10, 12, 18, 21, and 25). Humans, male and female, were created "in the image of God" (Gen. 1:27), and the initial result was "very good" (Gen. 1:31). The next chapter fills in some of the details that are especially rich from the perspective of a farmer. Much of my understanding about this perspective was gained from the theological works of Walter Brueggemann[3] and Theodore Hiebert.[4] In Genesis 2, God is presented as a gardener or farmer:

> And the LORD God planted a garden in Eden, in the east. (Gen. 2:8a, ESV)

This is an interesting view of one aspect of "the image of God." God's interest in mankind as agriculturalists is amplified by the fact that God made humans (in the Hebrew language, *'adam*) from the "dust of the ground" (Gen. 2:7). This dust comes from *'adamah*, the soil. This is a special kind of material that contrasts with *'erets*, the general term for dry land or earth used in Genesis 1:10. The *'adamah*, the material of *'adam*, is not just any old ground. It is the soil of arable cropland; it is good farming ground. The story goes on to show that God, the Farmer, made man to be a farmer:

> The LORD God took the man and put him in the garden of Eden to *work* it and *keep* it. (Gen. 2:15, ESV, emphasis added)

Other English versions translate God's purpose for man as tending, cultivating, and farming and as protecting, guarding, and looking after the garden.

Other rich insights that relate to the ancient structure and sociology of farming also are suggested in Genesis 2, but those points are for other chapters. The goal here is to get to the biblical picture of stewardship. Two verses are particularly important. They have been contrasted and compared by many writers delving into the religious components of either the abuse or care of the environment. Taking either alone leads to an unbalanced view of the first biblical instructions to humans. The first reads as follows:

> And God blessed them [the first man and woman]. And God said to them, "Be fruitful and multiply and fill the earth and subdue [*kabash*] it and have dominion [*radah*] over the fish of the sea and over the birds of the heavens and over every living thing that moves on the earth." (Gen. 1:28, ESV)

The transliterated Hebrew terms in italics are especially important for our purposes here. The term *kabash* means to subdue, to subject, to bring under control. A forcefulness, and even violence, is implied. The term *radah* means to rule, to take dominion, to prevail against. This passage is sometimes called "the dominion mandate" because the King James translation rendered *radah* as "have dominion over."

The critics of the Judeo-Christian tradition of environmental management have argued that this verse gives license to abuse nature. The paper by Lynn White[5] is a famous contribution to that argument. A contrasting view is that the verse is just a realistic assessment or description of human experience with wild nature. Wild nature resists gardening and farming, a point humorously made in a book by Michael Pollan.[6] The Bible does not romanticize or idolize nature. True to human experience through most of history, the Bible recognizes that there are thorns and thistles that invade the fields (Gen. 3:18), lions, bears, and wolves that attack the flocks (1 Sam. 17:34, John 10:12), and famines and plagues always threatening (Gen. 12:10, Ruth 1:1, Luke 15:14, and elsewhere). Whatever else is implied here, the real conflict between humans and nature is acknowledged, and humans are told to "have dominion over" nature. This has been strong motivation for the development of the medical and agricultural sciences. Only recently in human history have some people had the luxury of forgetting that nature needs to be subdued if humans are to survive.

All of this does not mean that the Bible gives us license to abuse our environment. That is a misinterpretation corrected by a second passage in the Creation story. We have already referred to Genesis 2:15. Here it is with key Hebrew terms added to amplify its meaning:

> The LORD God took the man and put him in the garden of Eden to work [*abad*] it and keep [*shamar*] it. (ESV)

The term *abad* also means to serve or cultivate. The implication here is that the human garden, that part of the earth we use, is to be served. That service takes the form of cultivating so as to realize the second objective of *shamar*. In addition to keeping, *shamar* means to care, guard, and protect, as in other verses where God cares for his beloved and we protect our eyes (Num. 6:24, Ps. 17:8, 91:11).

Taken together, the relationship of stewardship begins to unfold. God owns everything because he made everything. The retelling of the Creation story in the New Testament affirms this:

> All things were made through him [God], and without him was not any thing made that was made. (John 1:3, ESV; see also Col. 1:16)

God also controlled where humans were to live and what they were to do (Acts 17:26, Prov. 16:3). Initially he put the first man in a garden to have careful dominion, to serve it and rule it. Humans, made in the image of God, were to be servant-rulers. Jesus, the servant King, is the New Testament model of that concept.

The Genesis story of the Creation ends on a sad but hopeful note. Genesis 3 relates how humans became estranged from God. They ate "forbidden fruit" and broke the single rule that God had given them. In doing so, they damaged or broke all of their relationships. As a result, the ground ('*adamah*) was cursed (Gen. 3:17), and '*adam*, the human race, had to obtain food from it by toil and sweat (Gen. 3:19). In estrangement, humans became mortal, and in physical death, they returned to '*adamah*, the ground from which they were made (Gen. 3:19).[7] To make matters worse, humanity distorted God's purpose in Genesis 1:28, where they were told to "fill the earth." Instead of filling the earth as ecological servant-rulers, they filled the earth with wickedness (Gen. 6:5). God then unveiled the beginning of a plan to restore what was lost. He started over with the family of Noah (Gen. 6–11). In doing so, God restated the initial mandate:

> [1]And God blessed Noah and his sons and said to them, "Be fruitful and multiply and fill the earth. [2]The fear of you and the dread of you shall be upon every beast of the earth and upon every bird of the heavens, upon everything that creeps on the ground and all the fish of the sea. Into your hand they are delivered. [3]Every moving thing that lives shall be food for you. And as I gave you the green plants, I give you everything." (Gen. 9:1–3, ESV)

The expansion of diet is significant: Humans were redefined as potential predators. Beyond the simple ecological meaning is a connotation of something dangerous, something to be feared and dreaded by the rest of creation. Though this was a new beginning, God was working with a flawed, estranged humanity. He did not repeat the charge to "subdue and have dominion." The role of ruling was not reiterated. Instead there was a description of the damaged relationship that placed animals and ecosystems, and indeed all of the earth, into the hands of humanity, a dangerous humanity with the power to harm what they were supposed to serve. The mandate to serve (*abad*) and protect (*shamar*) (Gen. 2:15) stood unchanged, but with added clear application to all of the earth. At the same time, creation was left waiting and groaning for the fulfillment of God's plan of restoration (Rom. 8:19, 22). The remainder of the Bible can be summarized as the development of God's plan to bring all things back into their original right relationships (Col. 1:20; see also Isa. 11:7, 65:25; 2 Cor. 5:18).

## Biblical Stewardship

Though humans, including humans of all religions, have repeatedly damaged nature and *'adamah* and *'adam* by the improper use of the control they seized in breaking their intended relationship with God (Gen. 3), it was not meant to be so. It was not meant to be so even during the process before the full restoration of God's intended relationships. In the process, humans were to practice stewardship, and the principles of stewardship are laid down and repeated throughout the Bible.

*The first principle of stewardship* is that the "The earth is the Lord's":

> The earth is the LORD's, and everything in it. The
> world and all its people belong to him. (Ps. 24:1, NLT)

In the Hebrew Bible (Old Testament) alone, this idea is repeated many times (Exod. 9:29, 19:5; Deut. 10:14; Job 41:11; Ps. 50:10–12, 89:11), and it is reaffirmed in the New Testament (1 Cor.

10:26). The earth is God's because he made it (Gen. 1:1; John 1:3) and it is his because he purchased it (1 Cor. 6:20; 2 Cor. 5:18–19) and sustains it (Col. 1:17; Heb. 1:3).

*The second principle of stewardship* is that, though the earth is the Lord's, he has given its care into the hands of the human race. That is clear from what has already been discussed, but the concept also is repeated elsewhere:

> [6]You have given him [mankind] dominion over the works of your hands; you have put all things under his feet, [7]all sheep and oxen, and also the beasts of the field, [8]the birds of the heavens, and the fish of the sea, whatever passes along the paths of the seas. (Ps. 8:6–8, ESV)

The New Living Translation of verse 6 makes it very clear:

> You put us in charge of everything you made, giving us authority over all things. (Ps. 8:6, NLT)

The word for "dominion," or for being "in charge," used here, is not from *radah*, as in Gen. 1:28, but from *mashal*, often implying the rule of someone who is under higher authority. The principle of delegated human rule of the earth is repeated in another psalm:

> The heavens belong to the LORD, but he has given the earth to all humanity. (Ps. 115:16, NLT)

*The third principle of stewardship* is that stewards are responsible for the service, cultivation, protection, and care they provide. The first biblical example is in the Creation story, outlined earlier. In Genesis 3:9–13, God asks the first man and woman to give an accounting for what they have done, and in Genesis 3:14–25 the consequences of their misdeeds are spelled out. God expects humans to give good care to creation and not to waste or spoil the natural bounty:

> [18]Is it not enough for you to feed on the good pasture, that you must tread down with your feet the rest of your pasture; and to drink of clear water, that you must

muddy the rest of the water with your feet? [19]And must my sheep eat what you have trodden with your feet, and drink what you have muddied with your feet? (Ezek. 34:18–19, ESV)

God's judgment of human stewardship is both comprehensive and specific:

God will judge us for everything we do, including every secret thing, whether good or bad. (Eccles. 12:14, NLT)

[21]For only the upright will live in the land, and those who have integrity will remain in it. [22]But the wicked will be removed from the land, and the treacherous will be destroyed. (Prov. 2:21–22, NLT)

In a similar vein, Jesus promises the earth to the good stewards:

Blessed are the meek, for they shall inherit the earth. (Matt. 5:5, ESV)

Applying the concept of meekness to the trained obedience of a careful steward is consistent with its use in describing both Moses (Num. 12:3) and Jesus (Matt. 21:5, KJV).

Jesus told several parables that show accountability for one's stewardship. The evil farmers who did not practice good stewardship lost their leased vineyard and their lives (Matt. 21:33–45; Mark 12:1–12; Luke 20:9–19). The servant-steward who used his master's money wisely was rewarded, but the unwise servant-steward lost his position (Matt. 25:14–30; Luke 19:11–26). The shrewd steward was recognized for thinking ahead, even though he was dishonest (Luke 16:1–13).

The warning of coming judgment for what one has done with what one has been given is emphasized in the Christian scriptures. The strongest statement regarding stewardship of the earth is in the story of the last judgment at the end of the Bible:

The nations were angry with you [God], but now the time of your wrath has come. It is time to judge the

> dead and reward your servants. You will reward your
> prophets and your holy people, all who fear your name,
> from the least to the greatest. *And you will destroy all who
> have caused destruction on the earth.* (Rev. 11:18, NLT,
> emphasis added)

Another translation puts it this way:

> . . . [God's] wrath came . . . for destroying the destroy-
> ers of the earth. (ESV)

The judgment of a just God has both immediate and ultimate
consequences. Returning to the Creation story of Genesis, the
consequences of broken relationships mentioned earlier (Gen.
3:14–25) were not new punishments handed out by the Judge but
simple statements of what had already happened because of human failure. The Creation is so designed that how it is treated has
inherent consequences. The ground *'adamah* was cursed by *'adam's*
failure to keep their relationship with God and thus, indirectly,
with the ground. Everything is connected. God said, "Cursed is
the ground because of you" (Gen. 3:17, ESV), thus describing
what had already happened. Just the same today, failure of agricultural stewardship immediately depletes the soil, diminishes water
resources, and imperils the food supply. This is a sad and negative
situation, but one that has been addressed by faith. Faith's answer
is good stewardship of our natural and agricultural resources.
Stewardship makes humans responsible to serve (*abad*) and protect (*shamar*) their garden, a garden that now spreads wherever
humans go.

One possible interpretation of the cause of the curse, of the
character of the broken relationship with *'adamah,* is that it comes
from idolatry. The first couple desired the forbidden fruit as much
or more than they desired to maintain their relationship with God
(Gen. 3:1–6). When they ate the forbidden fruit, which grew from
*'adamah* as a part of nature, they started a "war with nature" (Gen.
3:17–19), manifested by weeds in their fields, wolves after their
flocks, and famine and plague around the corner. The general
human response to this situation has been to attempt to overpower
nature by control or to appease nature by worship, to play God or
to make idols. The biblical story also offers a peace treaty instituted

by God to end the war. The Ten Commandments (Exod. 20; Deut. 5) can be regarded as the conditions of such a peace treaty. The war with nature is addressed in the first two commandments:

> ³"Do not worship any other gods besides me. ⁴Do not make idols of any kind, whether in the shape of birds or animals or fish." (Exod. 20:3–4, NLT)

Playing God and having idols are forbidden. Part of the stewardship relationship is to return to God as God with nothing else in his place.

## Loving What God Loves

Even though approaching stewardship by consideration of negative consequences and corrective commands is a faithful representation of Scripture, there is another perspective that emphasizes the positive consequences of repaired relationships with God, nature, and fellow human beings. That work of reconciliation of "all things" is not complete, but it is well begun:

> ¹⁹For in him [Jesus Christ] all the fullness of God was pleased to dwell, ²⁰and through him to reconcile to himself *all things, whether on earth or in heaven*, making peace by the blood of his cross. (Col. 1:19–20, ESV, emphasis added; see also 2 Cor. 5:18)

This passage is so important to understanding God's love as the basis of stewardship that it bears repeating from another translation:

> ¹⁹For God in all his fullness was pleased to live in Christ, ²⁰and by him God reconciled *everything* to himself. He made peace with *everything in heaven and on earth* by means of his blood on the cross. (Col. 1:19–20, NLT, emphasis added)

This is the completion of the unfolding of the plan of restoration that God began with Noah. For those who hold the faith, the reconciliation worked by Christ means that they are now free to love what and who God loves. This is emphasized in another verse:

> For God so *loved the world*, that he gave his only Son, that whoever believes in him should not perish but have eternal life. (John 3:16, ESV, emphasis added; see also 1 John 4:9–12)

The word for "world" in the original Greek is transliterated as *kosmos,* which is the cosmos, all the universe, in English.

God's love for creation also is clear from his initial pronouncements that it was "good" and "very good." The point is repeated in several other biblical passages:

> The LORD is good to all, and his mercy is over all that he has made. (Ps. 145:9, ESV) [The New Life Version (NLV) says it like this: "... His loving is over all his works."]

> May the glory of the LORD last forever! The LORD rejoices in all he has made! (Ps. 104:31, NLT)

> You are worthy, O Lord our God, to receive glory and honor and power. *For you created everything, and it is for your pleasure that they exist and were created.* (Rev. 4:11, NLT, emphasis added)

Jesus Christ himself summarized the law and the prophets of the Hebrew Bible in just two commands, and both of them involve love: Love God totally and love your neighbor as yourself (see Matt. 22:36–39; Mark 12:29–31). This view of stewardship is one of love. We express our love of God and our love of neighbor when we give the creation good care. The application of this love is magnified when we recognize that the next generations are our neighbors in time.

Jesus came to make Christians his witnesses (Luke 24:48; Acts 1:8) and to give them fullness of joy (John 15:11, 16:24) and abundant life (John 10:10). When Christians fail to care for creation, they damage their witness. But when they live out the love that they have received and love what God loves with practical stewardship of creation, they are blessed with the joy of restored relationships with God, with neighbor, and with the rest of creation. That is abundant life.

# Chapter Three

## Ecology in the Bible

### Living in the Central Flyway

How does one acquire the heart of an ecologist? From my experience, it may help to live in a flyway. A flyway is the migratory path for North American birds between the warmer overwintering zones and the northern nesting regions. I grew up under the Central Flyway, an aerial highway for migratory waterfowl that fly to Canadian nesting grounds every spring and back to warmer climates each autumn. My home was situated on the "prairie plain" of north central Nebraska amidst rolling meadowlands that flooded every spring from a high water table. In March and April perhaps a quarter of our immediate landscape would be under seasonal spring pools. These pools served as resting and feeding stopovers for countless migratory ducks, geese, and sandhill cranes.

As a young boy, one of my favorite after-school activities in the spring was to scout for the migrating ducks. There was a springtime pond just north of a grove of trees that guarded our farmstead. Between the grove and the pond was a little hill, and I could crawl close to the ground to the high point and get within 100 feet of any ducks on the pond. That was close enough to get a good look at them before they would fly away at the sight of a human so near. There were mallards, pintails, canvasbacks . . . and sometimes blue-winged teal, the only species that stayed for the summer around the more permanent lakes in the region. I wished

29

that the ducks were less skittish, but they must have remembered the danger from hunters who also congregated in our area during the autumn migration. Usually the ducks would fly only 100 yards or so to the next pond, but I could not get closer than my first approach hidden behind the little hill.

It was an exciting day when my father gave me permission to take his newly purchased binoculars on one of my duck scouting expeditions. Even at 100 yards I could make out the green head of a male mallard. I spent more time than usual watching the ducks that evening because there was so much to see. Just as the sun went down and the cloudless sky was filled with a warm afterglow, I lifted the binoculars and looked west. As I focused for greatest distance an almost breathtaking sight became clear. There were uncountable migrating birds in the sky. Perhaps I saw 100,000 as I scanned the western view. There were ducks, but also Canadian geese and cranes. I do not believe I have ever seen so many living things at once. Some were still headed resolutely north, but many were circling, probably selecting an evening resting place. The sight was certainly unforgettable.

Years later I had a business meeting that sent me to Omaha during the time of the spring migration. I took the opportunity to make the four-hour automobile trip out to the ranch and visit my family who still lived on the "home place." It was perhaps thirty years since I had been there when the ducks were migrating. The spring ponds filled all of the familiar places, but there were very few ducks to be seen, so I asked my father if duck numbers had decreased over the years. He replied that there certainly were fewer migrating than there used to be. Together we wondered why. Was it pesticides, some other change in agriculture, or unfavorable disturbances in the wintering or nesting grounds? We were not sure. That evening, just as the sun went down, I borrowed my father's binoculars again and looked west into an almost empty sky.

I left with the sense that environmental deterioration had come very close to me that day. Perhaps because of my love for nature nurtured by migrating ducks, I had become an applied ecologist and spent a good part of my early career working to reduce the use of pesticides by developing methods of integrated pest management. That was a good thing to do, but still I had a hollow feeling on the inside. It was not enough. I needed to be even more focused on studying and communicating so that it made a bigger difference in our home, on this earth where we live. What God loves, and what I love, was showing signs of injury, and I needed to be even more involved in finding a sustainable means of recovery and maintenance. It was about that time that opportunities arose for me to become more involved in a new emphasis called "sustainable agriculture." Beginning with the personal glimpse I have just shared, this chapter is a journey across the fields of ecology and agricultural sustainability placed in the context of a biblical worldview that will hopefully inspire better environmental care.

## Holistic Thinking

One is not likely to get very far into a study of agricultural sustainability until one encounters a discussion of holistic agriculture. Thinking holistically about agriculture can be somewhat confusing, because it brings up a paradox. Depending on perspective, agriculture is both more inclusive than the natural sciences and also just one component of the natural sciences. The paradox is real because agriculture has social, cultural, economic, and religious components as well as aspects related to science, engineering, and technology. Thus the study of agriculture in the broad sense encompasses the study of the natural sciences. At the same time, agriculture occupies a place on the natural and scientific landscape that is embedded in nature. In the narrow sense, agriculture is just one slice of the pie that represents the natural sciences. This paradox is an example of the challenge to conceptual integration posed by the study of agricultural sustainability. Exploring religious concepts helps bridge the gap between the broad view and the narrow view of agriculture. The Bible is the main reference used here, but readers of other religions and value systems are invited to consider how they would link their worldview to the practical aspects of environmental care and food production.

Starting on the narrow side, agricultural sustainability is sometimes presented as the application of ecology to agriculture, especially the application of ecologically sound principles derived from studying natural ecosystems.[1] Old-fashioned crop ecology (my particular specialization) has evolved to include the study of "ecological agriculture" and "agroecology," complete with text-books,[2,3,4] and recently there has been an emphasis on reconnecting farms with natural ecosystems.[5] The foundational science for these works is of course ecology, defined as that branch of biology related to the study of living things and the environments where they live. Ecologists and environmentalists have become identified with the special-interest group concerned about the future because of past and present abuses of the natural world. Often they have sharply criticized Christian and Jewish rationalizations for environmental abuse, so an examination of ecology in the Bible is not introducing a new subject. The subject is complex because alternative interpretations are available. The main point here is that religious views do have import for ecology and the way we practice agriculture as we produce our food.

## The Christian Environmentalist

Not only has the rhetoric and deeds of a few environmental activists alienated many Christians, they also have alienated many farmers and people in the agricultural industries who work with farmers. Thus when I state that I am a Christian environmentalist, the response is sometimes wonder at the perceived oxymoron and suspicion regarding my good sense. A Christian farmer once advised me, "Do not be an environmental activist. Be an active environmentalist." So I have chosen this label for myself with some care. It is important at the outset to note that I have approached scriptural interpretation with the perspective of an ecologist or environmentalist, but also as one who believes that the scriptures are true and reliable when they are interpreted with reference to the whole of the Bible. In other words, passages of the Bible should not be explained in isolation but in harmony with all biblical teachings taken together. The ecological teachings of the Bible should thus be in harmony with the stewardship principles of the Bible laid out in the previous chapter. Though some of my thoughts about ecology in the Bible are unique, they should stand or fall based on the test of harmony. I share them because I believe they will stand.[6]

## Genesis 1

One of the most controversial and important parts of the Bible is the story of the beginning of all things as it is laid out in the first nine chapters of the first book. It is controversial because it does not match up with some aspects of the knowledge and theory of modern science. It is important because it summarizes the foundational understanding of Jews and Christians regarding their relationship to deity, to one another, and to the Creation. The Creation encompasses what is sometimes called the "cosmos." One of the main difficulties with interpreting the first part of the Bible is that it is not quite certain what it is supposed to be. It is obviously not a detailed scientific account. Indeed, detailed scientific vocabulary was not available when it was written. Still, some believe that it is literally true, while others dismiss it as mere ancient myth. Some take it as a kind of allegory that presents basic truth about the human condition. I have read that it is Hebrew poetry and have seen studies to show that it is not. For the believer, it is certainly a liturgy of worship. For a Christian environmentalist, it contains great ecological insight. Though some details of the story are contradicted by current scientific insight (e.g., plants were created in Gen. 1:9–13 before the sun and the moon in Gen. 1:14–19), with an ecological perspective, it can make a lot of sense.

The story of the Creation is partitioned into seven "days." That these are not necessarily literal twenty-four-hour days is clear from other passages:

> . . . one day is as a thousand years to the Lord, and a thousand years is as one day. (2 Pet. 3:8, referring to Ps. 90:4, ESV)

> . . . An entire lifetime is just a moment to you [LORD]. (Ps. 39:5, NLT)

On the first day of the Creation (Gen. 1:2–5), "God said, 'Let there be light,' and there was light" (ESV). From the ecological perspective, God first created energy and power. Psalm 62:11 confirms that power belongs to God. The modern big bang theory also predicts that light was the first creation.[7] On the second day (Gen. 1:6–8), "God said, 'Let there be space between the waters, to separate water from water' " (verse 6, NLT; instead of "space," the ESV has "expanse"). Realizing that there was not a scientific vocabulary to exactly describe what occurred, my interpretation is that on the second day, God created matter and space. On the third day (Gen. 1:9–13), the Bible says that God created the sea, the dry land, and the plants. With an ecological perspective, that can be interpreted to mean that God created environments, both aquatic and terrestrial, with photosynthetic organisms at the base of each food chain. On the fourth day (Gen. 1:14–19), God created "lights in the expanse of the heavens" (verse 14, ESV), the sun, the moon, and the stars. He made them "to rule over the day and over the night, and to separate the light from the darkness" (verse 18, ESV; see also Ps. 104:19). From the perspective of physical cosmology, the sun, the moon, and the stars would have been created before the plants, but from the perspective of ecology, this is about the seasons that define different kinds of environments: environments in general on day three and specific seasonal environments on day four.

The fifth day, described in Genesis 1:20–23, has special importance for the ecological perspective:

> [20]And God said, "Let the waters swarm with swarms of living creatures, and let birds fly above the earth across the expanse of the heavens." [21]So God created the great sea creatures and every living creature that moves, with which the waters swarm, according to their kinds, and every winged bird according to its kind. And God saw that it was good. [22]And God blessed them, saying, *"Be fruitful and multiply and fill the waters in the seas, and let birds multiply on the earth."* [23]And there was evening and there was morning, the fifth day. (ESV, emphasis added)

On the fifth day, the first blessing was pronounced by God, and it was on the fish of the sea and the birds of the sky. It was a blessing to "increase in number and fill" their ecosystems. Here is the introduction to what would later be called "population ecology." In the Creation story, the fish of the sea and the birds of the sky come before humans. It is significant to note that the ecosystems containing the fish and the birds are the first-named indicators of ecological health, with health meaning correspondence to divine blessing and intention. The psalmist agrees:

> [1]The earth is the LORD's and the fullness thereof, the world and those who dwell therein, [2]for *he has founded it upon the seas and established it upon the rivers.* (Ps. 24:1–2, ESV, emphasis added)

For an ecologist, this says that the stability of the biosphere rests on aquatic ecosystems. With regard to the birds, the prophet Jeremiah catalogued the reasons God destroyed the nation of Judah (chapters 1–29). One of the reasons was their failure to provide good ecological care for their land as commanded by God (Jer. 2:7, 9:10–14, 12:10–13). In describing the resulting devastation, Jeremiah noted that "All the birds of the air had fled" (Jer. 4:25, ESV). The Bible thus points to aquatic and avian health as a general test of conformity with God's environmental intention. In fact, Genesis 1:28 (following) and Psalm 8:6–8 make it clear that humans are, among other things, to be keeping track of the health of the fish and the birds.

On the sixth day (Gen. 1:24–31), God made the wild animals, livestock, and all of the creatures that move on the ground. Finally, to complete the process of making animals, he created humans in his own image, male and female (Gen. 1:26–27). God then pronounced the second blessing of the Bible:

> And God blessed them [the humans]. And God said to them, *"Be fruitful and multiply and fill the earth* and subdue it and have dominion over the fish of the sea and over the birds of the heavens and over every living thing that moves on the earth." (Gen. 1:28, ESV, emphasis added)

The blessing was repeated to Noah and his sons following the depopulation caused by the Flood:

> And God blessed Noah and his sons and said to them, *"Be fruitful and multiply and fill the earth."* (Gen. 9:1, ESV, emphasis added)

The first part of the blessing is just like that given to the fish and the birds. It is an ecological blessing, and thus fullness must mean an ecological fullness. The meaning of subduing and having dominion in Genesis 1:28 is detailed in the previous chapter on stewardship, but here it must be noted that humans also are given a responsibility to God, unlike the birds and fish. They are accountable to God not to "overfill" the earth and thereby to cause destruction. Again, the Bible does not give the details to satisfy an ecologist. It is not an ecological textbook. But it does point to examples and provide information about population options for humans.

Before examining in more detail what the Bible has to say about the population problem in the next section, the story of the days of the Creation should be closed by noting that by the seventh and last day, God had completed unveiling his plan for making everything (Gen. 2:1–3).

## The Population Problem

Overpopulation as an environmental crisis is first described in the Bible as it is related to livestock. Abram (Abraham) and Lot had more cattle than their grazing lands could support, so as good stewards they separated their herds and went to different grazing areas (Gen. 13). That is an example of simple, good management. However, just a few hundred years later, Abraham's descendants filled the land of Egypt:

> But their descendants had many children and grand-children. In fact, they multiplied so quickly that they soon filled the land. (Exod. 1:7, NLT)

To gain control of their population, the Pharaoh of Egypt imposed male infanticide, and that is generally regarded as an undesirable means of population management.

When it comes to the humans, the options for population control are controversial, especially as they relate to birth control and human abortion. One clear biblical option is celibacy, but that is clearly not for everyone (1 Cor. 7). Large families are described and blessed in the Bible, for example, the twelve sons of Ishmael (Gen. 17:20) and the one daughter and twelve sons of Jacob (Gen. 30:21, 35:22). In contrast, Abraham and Sarah (Gen. 17:19) were blessed with just one child, and Isaac and Rebekah (Gen. 25:24) with just two children. The Bible also pronounces blessings on the childless (Ps. 113:9; Isa. 54:1, 56:4–5; Gal. 4:27).

Since family size is not a biblical basis for pleasing God, it can be decreased when the land approaches "fullness." The challenge is to define and recognize such fullness. The condition of the birds and the fish, mentioned earlier, is a biblical guideline for detecting when humans are overflowing their ecological space. It also is significant that, according to the Bible, the state of fullness was reached in some places a long time ago (see aforementioned Exod. 1:7). Slavery, famine, plague, and war have reduced human population many times and are frequently mentioned in the historical parts of the biblical record. War in particular is a great destroyer of land and agriculture (Joel 1:6–12). But there are alternatives. Humans are told to seek knowledge (Prov. 2:3, 10:14, 15:14, 18:15), which can increase food production, protect and repair the environment, and care for human reproduction so that present and future people can be blessed with the abundance of creation. A full environment is a dynamic concept depending on the application of human wisdom and knowledge. It is clear that the land is "overfull" when the food supply is inadequate and the environment is deteriorating. The same God who declares children are a reward (Ps. 127:3) is also displeased when overpopulation destroys the environment (Jer. 2:7, 9; Rev. 11:18; also see Chapter 2).

## Humans and Nature

Many ecologists have questions not only about the Christian view of human population growth but also about the place of humans in nature. The concern is that the Bible may teach that as rulers of nature, humans are separate from nature.[8] Persons with that viewpoint might be more likely to abuse nature. Genesis 2 addresses that concern.

The Creation story is retold in the second chapter of the Bible, switching from the general perspective of the Creator in the first chapter to the specific viewpoint of the created humans in the second. Since the prior creation of animals (Gen. 1) was not experienced by the humans, in this story the animals are introduced after the creation of man. In addition, the creation of man and woman is separated in the second account:

> [T]hen the LORD God formed the man of dust from the ground and breathed into his nostrils the breath of life, and the man became a living creature. (Gen. 2:7, ESV)

In verses 2:18–22, the story continues first with the creation of the other animals and finally with the creation of the first woman from the side (rib) of the man. For our purposes here, it is important to see that the stuff for making the man was "the ground," in Hebrew, *'adamah*. The man, Adam, became a *nephesh*, "a living creature." When the animals were created, they had much in common with the man:

> So out of the ground ['*adamah*] the LORD God formed [or had formed] every beast of the field and every bird of the heavens and brought them to the man to see what he would call them. And whatever the man called every living creature [*nephesh*], that was its name. (Gen. 2:19, ESV)

The animals are made from exactly the same stuff (*'adamah*) and are the same kind of living creature (*nephesh*) as humans. The similarity of humans and animals can be missed in most English translations that say a man is "a living being" and an animal is a "living creature," but in the original Hebrew (and the ESV), they are exactly the same. The similarity of humans and other animals even extends to their spirit or breath:

> For what happens to the children of man and what happens to the beasts is the same; as one dies, so dies the other. They all have the same breath [*ruwach*], and man has no advantage over the beasts. (Eccles. 3:19, ESV)

The Hebrew word *ruwach* means breath, wind, or spirit.

The Bible is clear that humanity is a part of nature, not separated from it. The same physical and life principles work in animals and humans. However, the Bible does teach that there is a difference between humans and the rest of creation. Only humans are made "in the image of God" (Gen. 1:26–27). That uniqueness allows God to hold humans responsible to him for their care of the rest of creation (Gen. 2:15). Though aspects of physical death for humans and animals are the same, Christians believe that humans will be raised again for judgment and reward (Ps. 98:9; Isa. 11:4; Matt. 25:31–46; 2 Cor. 5:10; Rev. 20:11–15).

## Biological Diversity and Species Extinction

According to the Bible, God made everything there is (Gen. 1–2; John 1:3; Heb. 1:2). Creation reveals much of the nature of God (Rom. 1:20), and one aspect of that revelation is that God loves diversity. There are almost 20,000 known kinds of wild orchids, 19,000 kinds of asters, and 10,000 kinds of grasses.

> [24]O LORD, what a variety of things you have made! In wisdom you have made them all. The earth is full of your creatures. [25]Here is the ocean, vast and wide, teeming with life of every kind, both great and small. (Ps. 104:24–25, NLT)

> [10][God says] . . . all the animals of the forest are mine, and I own the cattle on a thousand hills. [11]Every bird of the mountains and all the animals of the field belong to me. (Ps. 50:10–11, NLT)

One purpose of a diverse nature is to declare the praise and glory of God:

> The heavens tell of the glory of God. The skies display his marvelous craftsmanship. (Ps. 19:1, NLT)

> Let the sea and everything in it shout his praise! Let the earth and all living things join in. (Ps. 98:7, NLT)

> [11]Let the heavens be glad, and let the earth rejoice! Let
> the sea and everything in it shout his praise! [12]Let the
> fields and their crops burst forth with joy! Let the trees
> of the forest rustle with praise. (Ps. 96:11–12, NLT)

Species extinction is a natural process, but species extinction accelerated by humans results in a decrease in the praise and glory otherwise belonging to God. The biblical pattern is for humans and the rest of creation to praise God together:

> All of your works will thank you, LORD, and your faith-
> ful followers will bless you. (Ps. 145:10, NLT)

God's faithful followers should therefore work to prevent species extinctions and the loss of diversity.

One of the clearest stories about God's care for the wild animals is that of the Flood (Gen. 6–9). The first human couple disobeyed God and broke their relationship with him (Gen. 3). Their descendants went from bad to worse and filled the earth with wickedness (Gen. 6:5). As the story goes, God started over again by destroying all of the people and animals on the earth with a great flood, but he spared Noah and his family. He instructed Noah how to build a great ark on which all of the kinds of animals also were saved and released to repopulate the earth. From this story, we know that God values both animals and people:

> Your righteousness is like the mighty mountains, your
> justice like the ocean depths. *You care for people and
> animals alike*, O LORD. (Ps. 36:6, NLT, emphasis added)

The book of Job (chapters 39–41) shows that God delights in creatures with no apparent value to humans.

The relative value of man and animals was addressed by Jesus:

> [11]He said to them, "Which one of you who has a sheep,
> if it falls into a pit on the Sabbath, will not take hold
> of it and lift it out? [12]Of how much more value is a
> man than a sheep! So it is lawful to do good on the

Sabbath." (Matt. 12:11–12, ESV; see also Matt. 6:26;
Luke 12:7, 24)

Thus there is a tension in the Bible regarding species extinction.
Wild and domesticated species are valued by God and serve and
praise him by their natural lives.[9] At the same time, people are
more valuable to him, and if a choice must be made, people have
priority. The answer to the dilemma is to be a steward of nature,
caring for the creation so that we do not "overfill" the land and
set up the dilemma.

## Other Ecological Issues

The Bible also addresses a few other ecological issues related to
wild nature. Following the Flood, Noah received this instruction:

> "Release all the animals and birds so they can breed and
> reproduce in great numbers." (Gen. 8:17, NLT)

Wildlife is to be given enough wild land so it can flourish. However, the hunting of wild animals is allowed (Gen. 9:2–3; Lev.
17:13; Deut. 12:15), as is the collection of the eggs of wild birds,
but the mother bird is to be protected (Deut. 22:6–7). The Bible
specifically allows the cutting of trees, provided they are not valuable for food (Deut. 20:20; 2 Kings 6:4; Luke 13:7–9). Forests may
be cleared for agriculture (Josh. 17:15, 18). Of course, the transition of wild land to agriculture is to be constrained by the needs
for wild species to prosper and for the soil to be protected.
Abraham is recorded as the first tree planter of the Bible (Gen.
21:33). Ezekiel 34:18–19 makes it clear that human impact on
nature should be minimized. Ecological succession or reversion to
a more wild state also is described in Scripture:

> Soon—and it will not be very long—the wilderness of
> Lebanon will be a fertile field once again. And the
> fertile fields will become a lush and fertile forest. (Isa.
> 29:17, NLT)

Psalm 104, one of the great nature songs of literature, praises wild nature and declares that it is "well cared for" by God (Ps. 104:16, NLT).

## Conclusions

In the beginning of this chapter, the focus was on the gap between the broad view and the narrow view of agriculture. It was suggested that a religious approach will help bridge that gap, will help restore a holism otherwise fragmented by compartmental thinking. The religious perspective is comprehensive, that is, life-encompassing, so it originates on the broad side of the divide. The narrow view that places agriculture in the context of wild nature has now been considered in some detail. It should be clear that the biblical or religious approach used here is interwoven with an understanding of wild nature and the ecological principles derived from the study of wild nature. Of course, one can take a compartmental view of ecology and leave the religious aspects out of the picture. But by bringing them in, an even greater level of holism is achieved, and many social, cultural, and ethical issues on the broad side are tied to the details of nature on the narrow side. Holistic or ecological thinking that includes the perspective of religion should help us see our environmental needs more completely and by touching the basis of motivation should help the person of faith act for the good of the environment.

# Chapter Four

## Land and Climate

### The Best Land in the World

Where would you go to find the best land in the world? To make the question just a little easier, where could you find nature's richest soils? We used to argue about that in my undergraduate soils courses at the University of Nebraska. It was a matter of regional pride to be an advocate for your own soils. Today, there are "state soils" designated for each state in the United States.[1] These are supposed to be representative soils instead of candidates for best soils, but it could be reasoned that given the climate, flora, and fauna of a region, these are the best soils. Still, if you wanted to find the soil that would grow the most food on the least area year after year, where would you look?

One could make strong cases for some of the flood plain soils in Egypt or Mesopotamia because they have supported civilizations for so long. Each year under natural flooding their fertility was renewed by deposits of nutrients and silt dropped from the Nile or the Tigris and Euphrates. One could argue that the soils along the Yellow and Yangtze rivers in China or along the Terai river valleys of South Asia must be very rich because of the massive human populations they support. An impressive set of soils could be identified on each continent. But I have to admit that I was really impressed with a soil that I saw just a few years ago.

43

I had been invited to co-teach a summer field course in the Midwest. Field trips and farm visits were the essential feature of the course. On one trip in Sioux County, Iowa, a young farmer took us out to see a soil pit, a hole in the ground that he had dug to study the rooting and water sources for his corn and soybeans. He stood in a hole that must have been almost six feet (about two m) deep. The soil was black grading into a very dark brown at about three feet. The farmer did not tell us the name of the soil, but the landscape supported very impressive crops. I could not make out a plow layer (a surface zone mixed by soil tillage) because the organic matter was so black and so deep. Thousands of years of prairie grassland had deposited masses of decaying roots throughout the soil profile to greater depths than could be seen in the soil pit. The roots were most abundant closer to the surface, and the black top layer appeared to be two or three feet thick. When I saw the soil in early August, the soybean roots could be seen four feet into the soil and still growing deeper, an insurance against a drought, which is common with shallower rooting zones. The climate of the area was usually moist enough that the lime layer, which marks the depth of typical water penetration in some grassland soils, was leached below the bottom of the pit. That meant there usually was enough rain and soil moisture to get a good crop. Yet the soil was famously rich because the rainfall was not so abundant that it washed away centuries of naturally weathered minerals needed for plant growth. I was impressed. Here was a candidate for the best soil in the world.

The next question to ask was how much is land with a soil such as that worth? The selling price at the time was somewhere between $3000 and $4000 per acre. That is not very much compared to a piece of downtown New York City. Yet with a little reflection, the reality of basic human needs should make such a soil priceless. Clean air, pure water, and good soil are worth more than gold. Because we have had them in abundance up until this time in history, we have taken them for granted and certainly undervalued them in our common economic accounting system. In reality, we could not live without them. Looking at that deep, black soil should inspire the awe of seeing a great cache of gold or a hoard of diamonds. Good land and a favorable climate are incredibly precious because of the part they play in sustaining life on earth.

If you have not thought about it before, it may not be clear why soil, land, and climate go together. They are familiar terms, but they have a range of meanings, so they must be defined for clear communication. The way the term is used here, land refers to the natural areas of the earth not covered by large bodies of water. It includes the soil and soil water, streams and ponds, plant cover, and other physical and biological characteristics, many of which are the result of climate. Climate is the average, seasonal pattern of weather in a region of the earth. Temperature and precipitation patterns are the two most important aspects of climate, and those patterns are determined by, among other things, "the lay of the land." That is, mountains, large bodies of water, elevation, and latitude are among the factors that shape climate. Thus land and climate are interacting dynamically, though slowly, and when we think about agricultural sustainability, it is helpful to think of land and climate together. The answers to questions about sufficiency of land to meet future demands for food depend on answers to questions about global climate change. Global climate is responding to land management. For example, tropical deforestation releases carbon dioxide to the atmosphere, and that promotes global warming. Perennial cropping promotes carbon accumulation in soils in the form of soil organic matter, and that reduces global warming.

## Good Land

Land is one of the key agricultural resources of the earth, and what farmers need and want is good agricultural land. Good agricultural land is the kind of land resource that can be used to produce food generation after generation without deterioration of its productive potential. First, it must be in a climatic zone that provides enough heat and precipitation so that crops and livestock can be productive. If the precipitation is too low, then it must be in a region where it can be irrigated. Second, it must have enough of the right kinds of soil so that the needs of crops and livestock can be met. Third, it must not be subject to severe erosion from wind or water when it is used for agriculture. Given these requirements, only about 11 percent (1.5 billion hectares, or 3.7 billion acres) of the ice-free land surface area of planet Earth qualifies as good agricultural land.[2] Another 13 percent could be used sustainably but with high economic costs for drainage, irrigation, fertility amendments, or intensive terracing to protect it from water erosion. Good land is constantly being lost to roads, urbanization, and deterioration caused by various kinds of bad management. In addition, the utilization of available land is not uniform around the earth. In the parts of Asia that are home to about 60 percent of the human population of the planet, essentially all good agricultural land is in use, and crop production has expanded onto environmentally fragile areas. In those places where that expansion is associated with poverty, severe land degradation is occurring. In contrast, good agricultural land is still available for agricultural expansion in Australasia, North America, and South America. The FAO estimated that, worldwide, 96 percent of all the good agricultural land was in use in 1993. With global climate change possibly affecting where crops can be grown, land and climate are critical concerns for the future.

## God's Land

The previous chapters on stewardship and ecology have laid down the principles of the Judeo-Christian worldview based on the premise that all land is God's land. The Christian believes that land is a gift from God. But it is neither a permanent gift nor a gift to waste. With land, God provides physical nourishment for creation. It is a means of blessing, not just for us, but for our neighbors in space and time (Lev. 19:18; Mark 12:31; Rom. 13:9). Our neighbors in time include future generations, and the Bible relates that God's intention (perhaps figurative) is to bless a thousand generations (Exod. 34:7; Deut. 7:9; Ps. 105:8). In any interpretation, the biblical view of God's concern stretches across time to include all people yet unborn.

In addition to land being a gift for the collective benefit of humans and other creatures, it also is viewed as a part of the symphony that is to praise and worship God:

The meadows are clothed with flocks of sheep, and the valleys are carpeted with grain. *They all shout and sing for joy!* (Ps. 65:13, NLT, emphasis added)

Let the rivers clap their hands in glee! Let the hills sing out their songs of joy. (Ps. 98:8, NLT)

Sing, O heavens, for the LORD has done this wondrous thing. Shout, O earth! Break forth into song, O mountains and forests and every tree! (Isa. 44:23, NLT; see also Isa. 49:13 and 55:12 for more singing mountains)

Of course, the valleys, hills, and mountains do not literally sing and clap their hands, but they do communicate a message:

³They speak without a sound or a word; their voice is silent in the skies; ⁴yet their message has gone out to all the earth, and their words to all the world. (Ps. 19:3–4, NLT)

Scriptural personification of the land also applies when it is damaged and ruined. The land is said to mourn (Job 31:38; Isa. 24:4, 33:9; Jer. 23:10; Hos. 4:3), and in anger, God removes law-breaking offenders from the land (2 Chron. 7:20; Hos. 9:15). Speaking specifically of the land promised to Abraham in ancient Israel and Judah (Gen. 12:7, 15:7, 17:8), God calls it "my land" in the following passage:

I will preserve a remnant of the people of Israel and of Judah to possess *my land*. Those I choose will inherit it and serve me there. (Isa. 65:9, NLT, emphasis added; see also Jer. 2:7, 16:8; Ezek. 36:5)

God's original instructions (Gen. 1–9) included land stewardship, and his blessing followed obedience and his judgment followed disobedience. Weather and climate were involved in the initial stages of blessing or judgment:

> [3]"If you walk in my statutes and observe my command-
> ments and do them, [4]then I will give you your *rains in
> their season*, and the land shall yield its increase, and
> the trees of the field shall yield their fruit. [5]Your thresh-
> ing shall last to the time of the grape harvest, and the
> grape harvest shall last to the time for sowing. And you
> shall eat your bread to the full and dwell in your land
> securely. (Lev. 26:3–5, ESV, emphasis added)

> [14]"But if you will not listen to me and will not do all
> these commandments, [15]if you spurn my statutes, and
> if your soul abhors my rules, so that you will not do all
> my commandments, but break my covenant, [16]then I
> will do this to you: . . . [19]and I will break the pride of
> your power, and I will make your heavens like iron and
> your earth like bronze. (Lev. 26:14–16,19, ESV)

The "heavens like iron" and "earth like bronze" are figurative descrip-
tions of drought. In a parallel passage, Deuteronomy says:

> The Lord will open to you his good treasury, the heav-
> ens, to give the rain to your land in its season. (Deut.
> 28:12, ESV)

But if you disobey,

> [23]. . . the heavens over your head shall be bronze, and
> the earth under you shall be iron. [24]The Lord will make
> the rain of your land powder. From heaven dust shall
> come down on you until you are destroyed. (Deut.
> 28:23–24, ESV)

Judgment also includes hot weather described as "fiery heat" in
Deuteronomy 28:22 and "fierce heat" in Revelation 16:9 (ESV).
Note, however, that such judgment did not come if some people
obedient to God remained in the land (Gen. 18:22–33), and God
"sends rain on the just and on the unjust" (Matt. 5:45, ESV).

Bad weather or climate change is just the beginning of judgment. Looking at biblical history, God allowed the ancients to go

only so far if they did not follow the instructions they received to care for his gift of land. One of the facts of history is that the nation of Judah and the city of Jerusalem were destroyed by Babylon in about 587 BCE. The prophet Jeremiah details the reasons for that catastrophe in the first twenty-nine chapters of the part of the Bible attributed to him. In abbreviated form, he taught that it was judgment caused by broken relationships. Those broken relationships were of three kinds: with God, with people, and with the land. Because the relationships are interrelated, they often are interwoven in the narrative and not easily separated; however, the careful reader will notice that the broken relationship with land is one reason for judgment. There are more examples, but three passages are sufficient to establish the point:

And I brought you into a plentiful land to enjoy its fruits and its good things. But when you came in, you defiled my land and made my heritage an abomination. (Jer. 2:7, ESV)

[10][The LORD says,] "I will take up weeping and wailing for the mountains, and a lamentation for the pastures of the wilderness, because they are laid waste so that no one passes through, and the lowing of cattle is not heard; both the birds of the air and the beasts have fled and are gone. . . ." [12]Who is the man so wise that he can understand this? . . . Why is the land ruined and laid waste like a wilderness, so that no one passes through? [13]And the Lord says: "Because they have forsaken my law that I set before them, and have not obeyed my voice or walked in accord with it." (Jer. 9:10, 12–13, EVS)

[10][The LORD says,"] Many shepherds have destroyed my vineyard; they have trampled down my portion; they have made my pleasant portion a desolate wilderness. [11]They have made it a desolation; desolate, it mourns to me. The whole land is made desolate, but *no man lays it to heart.* [12]Upon all the bare heights in the desert destroyers have come, for the sword of the Lord devours

from one end of the land to the other; no flesh has peace. ¹³They have sown wheat and have reaped thorns; they have tired themselves out but profit nothing. They shall be ashamed of their harvests because of the fierce anger of the Lord." (Jer. 12:10–13, ESV, emphasis added; see also Isa. 24:5–6)

Here is another translation of the key verse:

They have made it [the land] an empty wasteland; I hear its mournful cry. The whole land is desolate, and *no one even cares*. (Jer. 12:11, NLT, emphasis added)

When no one cared about the land, when no one cared for the land, the Bible and history record that the bad stewards were removed.

But that is not the end of the story. Full healing will come when all three relationships are restored. God promised the end of judgment with restoration for the people who would be faithful. In one of the great poems of all history, we read of God's plan:

He [God] will feed his flock [his faithful people] like a shepherd. He will carry the lambs in his arms, holding them close to his heart. He will gently lead the mother sheep with their young. (Isa. 40:11, NLT)

Once again, "The desert will blossom with flowers" (Isa. 35:1, NLT). Jeremiah, the same prophet who detailed the judgment, also foretold the restoration:

For behold, days are coming, declares the LORD, when I will restore the fortunes of my people, Israel and Judah, says the LORD, and I will bring them back to the land that I gave to their fathers, and they shall take possession of it." (Jer. 30:3, ESV; see also Jer. 29:10, 30:10, 46:27; Ezek. 36:28–38; Joel 2:18–27; Amos 9:13–15; Zech. 8:12)

The promised restoration also carried the hope that the restored people would be different, that God would write his law on their hearts:

But this is the covenant that I will make with the house of Israel after those days, declares the Lord: I will put my law within them, and I will write it on their hearts. And I will be their God, and they shall be my people. (Jer. 31:33, ESV; see also Ezek. 11:19 and 36:26 for the promise of a changed heart)

With a changed heart, God promises to heal the land:

Then if my people who are called by my name will humble themselves and pray and seek my face and turn from their wicked ways, I will hear from heaven and will forgive their sins and *heal their land*. (2 Chron. 7:14, NLT, emphasis added)

That changed heart also involves concern for future generations, for children:

[5]"Look, I am sending you the prophet Elijah before the great and dreadful day of the LORD arrives [the last judgment]. [6]His preaching will *turn the hearts of parents to their children*, and the hearts of children to their parents. Otherwise I will come and *strike the land* with a curse." (Mal. 4:5–6, NLT, emphasis added)

Land is a central concept of Jewish and Christian faith, and a comprehensive theological view has been developed for that perspective by Walter Brueggemann.[3] Our purposes are not so broad, but one aspect dovetails into practical applications in agriculture and needs to be mentioned here. In the law related to care of the land, the land was to receive a Sabbath rest.

## Sabbath Rest for the Land

One of the difficulties and mysteries of biblical interpretation is discerning what was specific for the ancient Jewish nation and what is intended for general applications. The Sabbath for the land is a case in point,[4] but some specific principles of stewardship can certainly be extracted and applied to modern concepts of crop rotation. The laws themselves are clear enough:

> [2]Speak to the people of Israel and say to them, When you come into the land that I give you, the land shall keep a Sabbath to the Lord. [3]For six years you shall sow your field, and for six years you shall prune your vineyard and gather in its fruits, [4]but in the seventh year there shall be a Sabbath of solemn rest for the land, a Sabbath to the Lord. You shall not sow your field or prune your vineyard. [5]You shall not reap what grows of itself in your harvest, or gather the grapes of your undressed vine. It shall be a year of solemn rest for the land. (Lev. 25:2–5, ESV)

The parallel passage in Exodus 23:10–11 makes it clear that spontaneous crops in the fallow period are for the poor and for the wild animals. Leviticus 25:18–22 says that the crops in the sixth year of the cycle will be sufficient for three years. Leviticus 26:34–35 and Leviticus 26:42–43 state that failure to follow this law will result in foreign conquest and deportation, so that "the land shall rest and enjoy its Sabbaths." Seventy years of exile from the land prophesied by Jeremiah (Jer. 29:10) provided the land with the rest that disobedient farmers had not given (2 Chron. 36:21). According to the biblical account, they had ignored seventy Sabbaths of the land.

From a religious perspective, this set of laws is clearly a measure of the faith of the ancient Jewish nation (and they did not measure up). From an agricultural perspective, this set of laws is about including a fallow period (a period with no crops) in the crop rotation (the sequence of crops grown on the land from year to year). The fact that the fallow or rest occurred only every seventh year presents an agricultural question. That is too infrequent to sustain the productivity of most agricultural land without substantial fertility inputs. The slash-and-burn system of land management works well provided that the land is given a long rest measured in decades. Many ancient farming systems probably involved rest every other year. The great agricultural advance of the Middle Ages in Europe was the "three-field system" that allowed cropping in two years out of three. Thus the Sabbath rest would sustain land only if it were tremendously productive and fertile, perhaps the best land in the world.

The underlying agricultural principle of Sabbath rest for the land is that there should be a fallow period in a crop rotation.[5] Continuous cropping violates that pattern. Much about this law appears to apply only to a specific people, place, and time in a specific relationship with God. The Sabbath law for the land is worth special note, though, because it underlines the biblical principle that God cares for the land, and he expects it to be well cared for.

## Then What Should We Do?

What are the practical applications of this biblical perspective? At least two must be emphasized. The first is that the biblical record does seem to give specific instructions for the care of a specific place, and similar instructions are certainly given by experience to others who settle down to care for a local land area. In my discussions with First Americans about their views on land stewardship, some shared that "first people" are given the first primary instructions for the care of a region, and they continue to hold responsibility for that care. It is not clear how those "instructions" are given, but certainly trial and error in primitive land management would lead to the accumulation of knowledge and wisdom that would eventually constitute a kind of home-field advantage in dealing with a set of local resources. This is an idea that shows up in the jargon of development as "local wisdom." Having a "sense of place" is a repeating theme in the literature on agricultural sustainability, a theme well developed in a good little book by Wes Jackson.[6] Jared Diamond's not-so-little book[7] presents the contrary view that technology will overcome local wisdom, but much of the concern about lack of sustainability may be linked to failure to heed local wisdom and of treating one place as if it were somewhere else.

Although most of the specific agricultural instructions in the Bible are in reference to a limited hill-farming area in the Middle East, the Bible does in fact note the principle of local wisdom as it could be applied anywhere:

> [18]And it is confirmed by the experience of wise men who have heard the same thing from their fathers, [19]those to whom the land was given long before any foreigners arrived. (Job 15:18–19, NLT)

Likewise,

> [14]For when Gentiles [non-Jewish people], who do not have the law, by nature do what the law requires, they are a law to themselves, even though they do not have the law. [15]They show that the work of the law is written on their hearts, while their conscience also bears witness. (Rom. 2:14–15, ESV)

In context, both passages are about more general concerns than agriculture, but agriculture is a specific application. The wonderful and special knowledge of a farmer is recognized in this passage:

> [24]Does he who plows for sowing plow continually? Does he continually open and harrow his ground? [25]When he has leveled its surface, does he not scatter dill, sow cumin, and put in wheat in rows and barley in its proper place, and emmer as the border? [26]For he is rightly instructed; his God teaches him. [27]Dill is not threshed with a threshing sledge, nor is a cart wheel rolled over cumin, but dill is beaten out with a stick, and cumin with a rod. [28]Does one crush grain for bread? No, he does not thresh it forever; when he drives his cart wheel over it with his horses, he does not crush it. (Isa. 28:24–28, ESV)

This kind of local knowledge is to be sought and respected (Prov. 19:27, 23:12; 1 Pet. 2:17; Rom. 13:7). A part of good land stewardship is respectfully receiving instruction from the people with local knowledge.[8] Wendell Berry is eloquent in describing how each field on a farm needs to be farmed with local knowledge that is ever-unfolding in the experience of a farmer who pays close attention to the land.[9] The consolidation of farms is causing the loss of this knowledge and threatens sustainability.

The second practical application of biblical land law has to do with a more general view of the principle of Sabbath rest for the land. In a functional, physical sense, one benefit of resting the land is to allow it to naturally recover from the adverse effects of all kinds of pollution. Today we are likely to think of land pollution in terms of toxic wastes added deliberately or inadvertently. However, the concept of pollution might be of biblical origin. Certainly the ancient texts use those ideas, but they stress that the source of pollution is law breaking. An example of law breaking is murder:

> [33]This [law] will ensure that the land where you live
> will not be polluted, for murder pollutes the land. . . .
> [34]You must not defile the land where you are going to
> live, for I live there myself. (Num. 35:33–34, NLT; see
> also Deut. 21:23b, 24:4b)

A clear implication is that modern pollution with its effects on land and climate is rooted in law breaking, including the law of love applied to future generations (see Mal. 4:5–6, quoted earlier). This is a natural rather than a man-made law, so the consequences cannot be avoided, even when the infraction is unintentional. Neither can such a law be revised.

Biblical land law graphically describes those consequences. The scriptures state that the land will vomit out those who pollute the land (Lev. 18:25, 28). Thus there is an inherent characteristic of the land that requires good care. Without good care, the capacity of the land to support people is lost. Biblically, land care is a matter of obeying God's natural laws. With the diversity of land in the world, such obedience involves the application of biblical principles in places that are not described in the Bible. It involves knowing land in specific local settings and caring for its needs. It is the sense of place and the reverence of place as God's creation. This sense of place fosters local adaptation of farming practices learned by the farmer in the process of farming.[10]

A very important general principle is that the land will care for us only if we care for the land. Soil nutrient depletion, accelerated erosion, desertification, salinization, and global climate change are just a few examples of responses of land to mismanagement. Of course, factors in addition to land management are involved in these problems, but the biblical picture is that the creation is so designed that mismanagement will be followed by bad consequences.

On the other hand, good management results in the opposite. A technical biblical term is righteousness. It simply means doing what is right:

> [17]And this righteousness will bring peace. Quietness and confidence will fill the land forever. [18]My people will live in safety, quietly at home. They will be at rest. [19]Even though the forest will be destroyed and the city torn down [by storms or by the development of changing civilizations], [20]God will greatly bless his people. Wherever they plant seed, bountiful crops will spring up. Their flocks and herds will graze in green pastures. (Isa. 32:17–20, NLT)

The Bible teaches that God loves the land and gives instructions on how to care for it. His followers should love the land too by keeping those instructions. Like the psalmist, they say, "Teach me your way, O LORD" (Ps. 86:11a, ESV). The lessons are ongoing and dynamic, based on a sense of place, a sense of land and climate, and the past, present, and future communities they support.

# Chapter Five

## Soil and Water

### Buried Fences

Soil stands out as particularly important to me because of what I have seen. It is a lasting but fragile resource. When European Americans killed the bison and plowed under the prairie vegetation on the Great Plains, the soil remained to support the new crops they tried to grow. The soil remains, or should remain, when we transform a natural landscape to an agricultural landscape. But the pioneers on the American Plains overextended their reach, and there have been repeated episodes of wind erosion and soil destruction. The "Dust Bowl" in the 1930s centered on the Great Plains of North America and resulted in the displacement of thousands of immigrant farmers who saw their hopes and dreams drift away as a drought cycle devastated an exposed soil resource. There are fields on my home ranch that were once plowed, then were blown by wind, and now have been reclaimed with native species as close to the original vegetation as possible. But every old field can still be seen. In the Dust Bowl days, the wind-blown soil drifted like winter snow, burying fences, but unlike the snow, the drifts have not melted away. There is still the tell-tale ridge where an old fence line collected blowing soil. But more to the point, there is reduced plant productivity where the rangeland was once plowed. After seventy years, the soils are still depleted, and the loss has not been recovered. This story can be retold for many different cases. In some places, damage from salinization (soil deterioration caused by salty irrigation water) has persisted for thousands of years. Many changes in the soil are lasting changes, so soil management is a critical issue for agriculture. And water for agriculture is so closely linked to soil that it is appropriate to consider soil and water together.

Soil is the foundation of the food system. It is subject to destructive change, and without soil, the world would starve. Even with modern knowledge, humans have not invented an alternative for soil that can function on a scale capable of supporting the massive amount of food production we now require. Agricultural theologian Mark Graham names the quality and quantity of topsoil as the most important criterion of health in an agricultural system.[1] Closely linked to the soil are the surface and groundwater (aquifer) supplies that are filtered and recharged as rainfall and irrigation pass over and through the soil.

Soil as a subject is something most people hardly think about, and when they do, it is usually phrased in prejudicial parlance about "dirt." Yet soil science is a rich and diverse topic divided into at least five subdisciplines: soil formation and classification, soil physics and water relations, soil chemistry and mineralogy, soil fertility and fertilization, and soil biology. Though mentioned only once, soil water is important in each area of soil science. Modern soil science has incorporated the various practical aspects of soil management across the subject spectrum.[2] In fact, integration across that spectrum needs to be stressed. Multiple college courses, whole departments, and several scientific journals have been devoted to the subject. My goal here is not to encapsulate the academic aspects of such a diverse subject, but my personal perspective leads me to two summary points based on the science and management of soil: (1) Soil is extremely important, so important that the natural order gives priority to the protection of soil. (2) Soil is extremely important, so important that every person should be concerned about its state of health. The repetition is deliberate. We cannot live without soil and the water resources that interact with soil.

With the perspective of an agronomist (and with the help of some theologians too), I can see references to soil and soil management in the Bible. Those references were probably introduced as familiar metaphors for the original audiences of the scriptures.

That familiarity is now largely lost on populations several generations removed from the practice of agriculture. My primary goal here is to point out and interpret those references. Hopefully this will facilitate appreciation and communication that respect and connect science and faith. That is required by any truly holistic approach to sustainability.

## Soil in the Bible

Is the claim that soil is vitally important to human survival consistent with the Bible's presentation about soil? I think it is, but it may take some interpretation to support the claim. As is true of most religious writings, the Bible is not primarily about the physical or material aspects of human experience and need. It instead focuses on the human experience of nonmaterial reality and the resulting implications of how one ought to live. Nevertheless, the Bible uses physical and material examples to help communicate understanding about nonmaterial reality, and it extends the applications of that understanding to encompass the physical and material aspects of human existence, including the practical care of our natural and social environment. Thus the information is there, but it often requires some explanation.

Beginning with the story of the Creation in the first part of the Bible, there are regular references to the soil and its significance to human life. These ideas have been developed by several theologians and environmental writers, but I particularly have been influenced by Brueggemann[3] and Hiebert.[4] They provide a foundation for understanding agricultural and environmental stewardship from a biblical perspective.

In the Creation account, we learn that God is the first gardener or farmer:

> And the Lord God planted a garden in Eden, in the
> east, and there he put the man whom he had formed.
> (Gen. 2:8, ESV; see also John 15:1–8 and 1 Cor. 3:7, 9)

As an aspect of being created in God's image (Gen. 1:26–27), humans were made to be gardeners and farmers:

> The Lord God took the man and put him in the gar-
> den of Eden to work it [cultivate or farm it] and keep
> it [care for it]. (Gen. 2:15, ESV)

As Hiebert[5] points out, the soil imprint on humanity is thorough-
going. In the Hebrew language, the Earth as dry land is 'erets (Gen.
1:10). Humanity ('adam) was created male and female in the image
of God (Gen. 1:27). When the details are given in the second
chapter of the Bible, the "stuff" of which 'adam (the human race)
is formed is "the dust of the ground" (Gen. 2:7). This dust comes
not from 'erets but from 'adamah. The 'adamah, the material for
'adam, is not just any old ground, it is the soil of arable cropland,
good farmland. Genesis 3 tells the story of human estrangement
from God. As a result, the ground ('adamah) is cursed (Gen. 3:17),
and 'adam must obtain their food from it by toil and sweat (Gen.
3:19). In estrangement, humans have become mortal, and in physical death, they return to 'adamah, the ground from which they were
made (Gen. 3:19). The Bible analogy is that we are made from the
same kind of soil that produces our food. With this interpretation,
the soil-food-life connection of the Bible is very strong, right down
to the rudiments of the nutrient cycle.

The distinction of various kinds of land and soil is repeated
throughout the Bible. Fertile soil is a blessing (Deut. 28:4). Bar-
ren fields with bad soil are a curse (Deut. 28:18). Isaiah (58:11)
refers to "a well-watered garden" as the example of blessed fruit-
fulness, and Ezekiel (17:8) speaks of "good soil" with "plenty of
water" in making a similar point. One of the repeated parables of
Jesus (Matt. 13:4–8; Mark 4:4–8; Luke 8:5–8) identifies different
kinds of soil: that which is hard in the footpath, that which is
shallow and rocky, that which is weedy with thorns, and that which
is fertile giving an increase of thirty, sixty, or even a hundred times
over what was planted.

The problems of soil erosion are certainly ancient, and they
are used in the Bible to illustrate despair and hopelessness. The
book of Proverbs (28:3b) refers to a beating (erosive) rain that
leaves no food. The idea is developed more fully in the book of
Job. Job tells his companions about the erosive effects of life's
troubles on human hope. He complains that those troubles are

allowed by God, and he compares the effect of those troubles to the erosion of rock and soil caused by water:

> [18]"But as mountains fall and crumble and as rocks fall from a cliff, [19]*as water wears away the stones and floods wash away the soil, so you* [God] *destroy people's hope.*" (Job 14:18–19, NLT, emphasis added)

Wind erosion also is mentioned in Scripture in conjunction with drought, crop failure, and starvation. It says that during drought, the rain shall be like powder, and that "dust shall come down on you until you are destroyed" (Deut. 28:24, ESV). Dust storms are generated by wind erosion.

Previous chapters have addressed the reasons for those troubles and what the Bible says must be done to avoid or stop them. In the present context, note that there is a vital connection between soil and food and life. For the ancients, soil loss meant loss of food and hope. In many situations, the same is true today.

## Water

The ancient Hebrew people lived in a climate and followed an agricultural system that depended on rainfall. They saw an intimate association of soil and water:

> [4]This is the account of the creation of the heavens and the earth. When the LORD God made the heavens and the earth, [5]there were no plants or grain growing on the earth, for the LORD God had not sent any rain. And no one was there to cultivate the soil. [6]But water came up out of the ground and watered all the land. (Gen. 2:4–6, NLT)

This moistened soil was used by God to form the very first man much as a potter would shape clay (Gen. 2:7). In the completed creation, soil and water are still perceived in relationship to agriculture:

> "The rain and snow come down from the heavens and
> stay on the ground to water the earth. They cause the
> grain to grow, producing seed for the farmer and bread
> for the hungry." (Isa. 55:10, NLT)

Isaiah goes on to say that the blessing and protection of God is "like a watered garden, like a spring of water, whose waters do not fail" (Isa. 58:11, ESV). Water as the catalyst for agricultural productivity also is used as a metaphor by the prophet Ezekiel:

> [The vine] had been planted on good soil by abundant
> waters, that it might produce branches and bear fruit
> and become a noble vine. (Ezek. 17:8, ESV)

Irrigation is mentioned in the Bible in the context of the Hebrew migration (the Exodus) from Egypt's irrigated Nile Valley to the land of Canaan that depended on rainfall:

> For the land that you are entering to take possession of
> it is not like the land of Egypt, from which you have
> come, where you sowed your seed and irrigated it, like
> a garden of vegetables. (Deut. 11:10, ESV)

This particular passage is doubly interesting, because it is one of the few places that vegetable production is mentioned in ancient writing. It indicates that home gardens may have been routinely irrigated, at least to assist the germination of seeds. The Scripture also describes the results of irrigation management:

> [33]He [God] turns rivers into a desert, springs of water
> into thirsty ground, [34]a fruitful land into a salty waste,
> because of the evil of its inhabitants. [35]He turns a desert
> into pools of water, a parched land into springs of water.
> (Psalm 107:33–35, ESV)

Verse 33 is probably a description of the failure of the flood of the Nile. Verse 34 is certainly a description of the result of poor irrigation management, a process we now call "salinization," which is the accumulation of salt in irrigated soils. Farmers and soil scientists have developed a whole set of practices to prevent or delay salinization, but those practices were unknown to ancient farmers and are unavailable for some modern farmers who are hard-pressed to maximize immediate food production. The "evil" cause of the salting of soil in verse 34 is simply poor stewardship of soil and water and oppressive economic systems that cause such poor stewardship to occur. Verse 35 can be interpreted as a reference to land reclamation in the desert. The prophet Isaiah also refers to this:

> [T]he burning sand shall become a pool, and the thirsty
> ground springs of water; in the haunt of jackals, where
> they lie down, the grass shall become reeds and rushes.
> (Isaiah 35:7, ESV)

Although we may think of irrigation management and land reclamation exclusively in terms of agronomy and engineering, the aforementioned passages show that the ancients recognized that God was involved. And this was especially true of rain:

> Ask rain from the Lord in the season of the spring
> rain, from the Lord who makes the storm clouds, and
> he will give them showers of rain, to everyone the veg-
> etation in the field. (Zech. 10:1, ESV)

Like soil, the ancient writings view water as very important. The biblical description of the original Garden of Eden stresses that it was well watered by a river (Gen. 2:10).

## What Is Essential?

Both soil and water are essential to life. Perhaps that statement is self-evident, but it is probably worthwhile to consider the lists of essentials in just a bit more detail. One way of looking at necessity or essentiality is in terms of how long we could live without a thing. On that basis, the first factor on a list would be breathable air. After that would be potable water, or possibly shelter, depending on the particular stresses of the environment. In warm or hot places, water would come before shelter, but in very cold places, shelter would be of greater urgency. Next on the list would be food, and assuming that the food provided all of the required nutrients for life, we have a complete list of physical needs. Of course, there also are psychological and social needs, but for now the physical list is enough to consider.

In light of the strong assertions made earlier about the importance of soil, it is striking that soil is not on the list. But soil is a great "ecological servant," and each item on the list of essentials is ecologically linked to soil. The linkage to food should be clear. The crops we eat and use to feed our livestock grow in soil and depend on soil as the source of the mineral elements they need. Likewise, many of the construction materials for shelter are made from soil or grow in soil. Water, especially pure water fit for drinking, also is linked to soil. Soil is a filter to remove contaminants in water and a storage location for the water used by land plants. Coming to the top of the list, breathable air also is linked to the soil. Many atmospheric pollutants settle on soil, and some are detoxified by soil microorganisms. Greenhouse gases also are removed by soil processes. In contrast, some soils release greenhouse gases to the atmosphere, but the biosphere as a whole is maintained in a condition amenable for life by giant ecosystem services carried out by oceans, forests, and soils. If the Bible had been written by people living in the frozen Arctic or in a food-rich oceanic environment, then the sea instead of the soil would probably have been the great metaphor for life. The key point is that the ancients knew that untamed nature is essential, and we are wise to know it too, especially when a modern lifestyle isolates many of us from the day-to-day associations with soil, forests, and oceans.

Of the four essentials just listed, food is probably the one item most completely dependent on soil. According to FAO data, only about 1 percent of the world's food supply comes from oceans, lakes, rivers, and streams. The remainder of our food comes from the practice of agriculture based on the soil. Thus when one considers agricultural sustainability, there is another list of essentials related to what it takes to have a wholesome, abundant, and lasting supply of food. It is the list of essentials for agriculture, and one of the goals of this chapter and of those that follow is to ascertain what is on that list. If something is a basic necessity, then we can expect it to show up in some way in the recorded wisdom of ancient people. When it is confirmed by modern knowledge and practice, we have probably identified something that is essential. Thus the biblical perspective that we have been examining will be a fruitful approach for detecting what is essential for agriculture.

## Essentials for Soil and Water

Soil and water are focal points of stewardship (Chapter 2) and ecological principles that apply to agriculture (Chapter 3). Soil and water also are just two of the natural resources encompassed by the words "land and climate" (Chapter 4). For example, land and climate also include the biodiversity of wild nature that provides ecosystem services such as natural pest control. In addition, natural resources are organized in unique local combinations, and when they are used in agriculture, they should be managed locally with ongoing refinement of practical knowledge that leads to conservation. All kinds of agriculture depend on many natural resources, so the first essential of agriculture simply reflects that general dependency.

- Adequate natural resources must be managed in locally appropriate ways.

For sustainability, natural resources must be respected, protected, and conserved. Stewardship of nature and the understanding and application of ecological wisdom are needed to maintain and sustain the natural resource base of agriculture.

Because soil occupies a central place at the foundation of the food system, it requires special attention.

- Soil must be respected and protected.

This simple statement summarizes the function of soil as the source or "ground" of life. Soil care equates to environmental and social well-being, but soil loss equates to loss of hope. Soil erosion, both from wind and water, soil salinization from faulty irrigation, and desertification from the failure of water supplies were all given as examples of failures in soil protection. Other factors also contribute to the damage or destruction of soils.[6] Nevertheless, the point for soil protection is well established. Good agricultural soil is a limited natural resource that is almost completely utilized today, so soil health and conservation are topics of increasing urgency.

Other very important points about soil should be made. One example is how soil is managed to provide a sustainable supply of nutrients for crops and livestock. However, those topics are aspects of how we do agriculture rather than of what we need for agriculture, so they will be addressed later when we consider farming and the practices of farming.

Water also stands out on the list of natural resources needed for agriculture. The challenge is to have the right amount and kind of water.

- Water for agriculture must be managed.

This statement is drawn more from agronomic understanding than from the ancient sources, but based on what we know today, passages of the Bible support this idea. You have to have water for crops to grow and for animals to drink. Sometimes the water comes from rain, in which case the soil must be protected from water erosion, as noted earlier. That means managing the water and the soil together with such practices as land terracing, sod waterways for water movement off of fields, and sod or wooded strips along streams. Sometimes the water comes from irrigation, and some irrigation water can contain harmful salts. There is a complex of knowledge and methods needed for sustainable irrigation from surface waters and water wells. And again it is integrated with soil management through the prevention of the problems of salinization and desertification.

Water management problems in agriculture also are associated with having too much water in the soil. This is water management by drainage. On a grand scale, it includes holding back the ocean with the dikes of the Netherlands. On a more modest scale, it involves the installation of subsurface drainage tiles that channel water into ditches or other constructions to carry excess water from the land. This is a topic that I have not seen mentioned explicitly in the Bible, probably because the authors came from relatively dry lands. However, the potential problem or opportunity is seen in passages such as Isaiah 35:7, quoted earlier. The passage describes the invasion of grassland by reeds and rushes. It might be possible by practices of reclamation to return grass to where reeds and rushes have taken over. Of course, there is another modern issue of environmental management here. What about wetland protection? Here the political processes of society work out the managed balance of wild biodiversity with needs for food and desires for economic development. It is not clear if the ancient agricultural systems based on wetlands (e.g., wetland rice in Asia) ever considered such a balance and what the consequences of the loss of wild prehistoric wetlands have been in many parts of the world.

## Conclusions

Our starting objective was to identify and interpret ancient metaphors that can be related to the appreciation and management of soil and water resources, the foundational resources for agriculture. The Bible in fact contains much relevant information. That information can be summarized in three essentials for agriculture:

- Adequate natural resources must be managed in locally appropriate ways.

- Soil must be respected and protected.

- Water for agriculture must be managed.

These are just the start of a list of factors that will be augmented in the following chapters (and see Appendix 3).

# Chapter Six

# Crops, Seeds, and Food

## Gardening Runs in the Family

I live in a two-story, 1930s house about one mile from the Cornell campus in Ithaca, New York. Its location has enabled me to walk back and forth to work with all of the advantages of the exercise and time spent in quiet reflection or discussion with my wife, who often walks with me. Another great advantage, which may not be immediately obvious, is that the nearness of home and work has enabled me to be a gardener. Gardens are best when there are a few minutes every day to do the seasonal tasks. This is especially true at the harvest season, when timeliness ensures the best of food freshness and quality. By living in a way that avoided uncertain and prolonged commuting, I have been able to spend a few moments each morning or evening doing what is necessary to keep my hobby going and growing.

In addition to the advantages of relaxation, exercise, and good food, being a gardener also maintains a family tradition. My great-great-grandfather immigrated to the Chicago area in 1868 from German-speaking East Pomerania. He became a truck gardener growing vegetables for the city market. His son-in-law (my great-grandfather) worked with him on their vegetable farm. His daughter (my grandmother) came to Nebraska in 1917 and married a rancher. She trained her children to grow vegetables, and that helped them survive the Great Depression of the 1930s, when they not only grew a garden for their own use but sold vegetables to neighbors as they could. All of her children, including my father, have grown vegetable gardens. I remember my father's seed collection kept in an assortment of small glass jars with tight lids. They were stored in a metal-lined drawer that protected them from mice that might get into the house. And my father taught me to be a gardener. I now have a small backyard vegetable garden plus a row of rhubarb and four blueberry bushes. There also are numerous plantings of ornamentals, including a small collection of roses that gives me great enjoyment.

My wife does not have an unbroken chain of vegetable growers for ancestors, but one grandmother in California delighted us with her love for the garden. She had no grass to mow because her whole yard was devoted to vegetables, flowers, and fruit trees. She advised us to always grow vegetables and flowers together because one fed the body and the other the soul. When we became parents, it was natural that our children should join me in the garden as soon as they could. I do not know if all of them inherited my love for gardening, but each one (we have three sons) especially delighted in collecting earthworms when I turned the garden soil with my fork each spring. I have a precious picture drawn by my youngest son when he was seven years old. It is a crude representation of the two of us in our big garden boots. He has a giant red earthworm in his hand and it says, "I love it when we plant in the garden and find big worms." Each of our sons helped me harvest fresh produce, and each has a special appreciation for food fresh from the garden. Probably the love for gardening has as much to do with training as it does with ancestry.

It is good to have a garden, and I sometimes wonder what our troubled world would be like if each person could grow some of her or his own food. I believe the prophet Micah must have thought about this too:

> [3]. . . they shall beat their swords into plowshares, and their spears into pruning hooks; nation shall not lift up sword against nation, neither shall they learn war anymore; [4]but *they shall sit every man under his vine and under his fig tree*, and no one shall make them afraid, for the mouth of the Lord of hosts has spoken. (Micah 4:3b–4, ESV, emphasis added)

Author Michael Pollan proposes this interesting thesis: The part of the earth under human care really is a garden and a school.[1] In the broadest sense, that garden also includes our field crops, pastures, and rangelands. That big garden certainly provides some of our most important education and most of our food.

Essentially all of our food comes directly or indirectly from plants, from organisms that transform sunlight energy into substances that we can digest. The most complex photosynthetic organisms reproduce with seeds. Archeologists and anthropologists generally agree that the discovery and invention of agriculture centered on human attention to the management of seeds. Seeds that were nutritious, that could be easily collected and stored, and that could be planted to grow more seed are one of the prehistoric foundations of agriculture. Although crops that could be grown without seed (e.g., potatoes and garlic) and vegetable gardens that conveniently provided favorite foods close to home may also have been important in the evolution of agriculture, early civilizations developed around crops with seeds. Most important were the cereal grasses. The Middle East and India depended on wheat and barley, East Asia on rice, the Americas on corn (maize), and Africa on sorghum and millet. These crops are now of worldwide importance. Wheat, rice, and corn provide more than two thirds of the total crop production of the modern world.[2] Most of the food for humans and the feed for livestock comes from a short list of crop species. Perhaps agricultural sustainability would be promoted with more diversity in food production, but our present food supply also is based on thousands of years of know-how that should not be ignored. Our goal here is to continue the examination of the ancient wisdom about farming, especially as it is recorded in the Bible. In considering the examples with crops, seeds, and food, we also will be looking for additions to our list of the essentials of agriculture. What are those necessities that farmers have always had to find as they grow our food?

## Kinds of Crops

From the beginning, in the biblical story of the Creation, plants are identified as the basis of our food system:

> And God said [to the first humans], "Behold, I have
> given you every plant yielding seed that is on the face
> of all the earth, and every tree with seed in its fruit.
> You shall have them for food." (Gen. 1:29, ESV)

Because the authors of the Bible lived in the Mediterranean zone,
the Bible does not mention the full array of crops now important
in the world. For example, wheat is mentioned forty-seven times
in the ESV, from the first book (Gen. 30:14) to the last (Rev.
18:13), and barley is mentioned thirty-five times; however, rice
and corn are not mentioned at all. (But note that in the once
popular King James translation, wheat is called "corn.") Lentils
(an important grain legume) are mentioned three times in the
Bible, and beans are mentioned twice. The book of Ezekiel has a
typical passage naming crops of the biblical world:

> ". . . take wheat and barley, beans and lentils, millet and
> emmer, and put them into a single vessel and make
> your bread from them." (Ezek. 4:9, ESV)

Millet and emmer are other kinds of cereal crops.
    A vegetable garden is mentioned once in the ESV transla-
tion (1 Kings 21:2). The exiled Hebrews are told to plant gardens,
presumably for food (Jer. 29:5). Vegetables are mentioned four
times. Herbs are mentioned five times, and some references are
probably to edible vegetables. The following is a typical reference
to those kinds of food:

> We remember the fish we ate in Egypt that cost noth-
> ing, the cucumbers, the melons, the leeks, the onions,
> and the garlic. (Numbers 11:5, ESV)

Onions and garlic often are grown from bulbs or cloves and do
not require seed.
    Among the vines and tree crops, the grape is very important.
It is named only twelve times, but wine, which is made from grape
juice, is recorded 216 times. Raisins (dried grapes) are mentioned
nine more times. The olive is mentioned thirty-seven times. Oil,
most often the product of the olive, is mentioned 197 times. The

date palm is mentioned more than thirty times, and the sugary syrup of the date fruit is usually translated as "honey" in the frequent references to the land of "milk and honey" (e.g., Exod. 3:8). The almond, apple, fig, and pomegranate are other biblical tree crops. Mint, dill, and cumin are among the biblical herbs and spices. A characteristic of all the biblical crops is their adaptation to the climatic zone where the biblical authors lived. They were familiar foods or condiments and thus useful models for biblical lessons.

If one turns to the First Americans, who had developed sophisticated agricultures, the crops represented in their religious traditions vary because of different adaptation and availability as they were domesticated. Important in the culinary and religious traditions of the first nations of eastern and central North America are the Three Sisters (corn, squash, and beans) plus sunflower farther to the west.[3] In the Andean region of South America, potato, a crop usually grown from tubers instead of seeds, would be added to this list.

## Seeds Are the Source

Although the major religious writings of the world may point to the pre-agricultural roots of the human race, the invention of writing came after the invention of agriculture, so all of the ancient writings come from authors who are familiar with agriculture. Even in the first chapter of the Bible, emphasis is placed on the fact that reproduction by seed was somehow important (Gen. 1:11). Reproduction by seed is the foundation of most kinds of crop production. Even garlic, onions, and potatoes produce seed, though usually their seed is not used to plant a crop. The Bible states explicitly that seed-bearing plants were given to humans for food (Gen. 1:29) and that "seedtime and harvest" are a part of the God-ordained pattern of life on earth (Gen. 8:22):

> They [a people blessed by God] sow their fields, plant their vineyards, and harvest their bumper crops. (Ps. 107:37, NLT)

Solomon, the king known for his wisdom, gave the following advice:

> In the morning sow your seed, and at evening with-
> hold not your hand [from sowing seed], for you do not
> know which will prosper, this or that, or whether both
> alike will be good. (Eccles. 11:6, ESV)

The biblical patriarch Joseph guided his adopted country of Egypt
through seven years of famine (Gen. 41–47). He wisely saved seed
for planting crops when the famine was over (Gen. 47:23–24).
Seed for sowing is also mentioned in the laws of ancient Israel
(Lev. 11:37, 26:16). Jesus told parables around the theme of planting seed (Matt. 13:3; Mark 4:3; and Luke 8:5). The Apostle Paul
also used seed to illustrate his understanding of how God works
with people:

> He who supplies seed to the sower and bread for food
> will supply and multiply your [spiritual] seed for sow-
> ing and increase the harvest of your righteousness.
> (2 Cor. 9:10, ESV)

In addition, Paul said, "Whatever one sows, that will he also reap"
(Gal. 6:7, ESV; see also Matt. 7:16–20), thus drawing a spiritual
principle from the biological understanding of the necessary linkage between the seed and the crop that follows.

The Bible also tells us that tree crops were planted and
highly valued. Leviticus 19:23 (ESV) begins, "When you come into
the land and plant any kind of tree for food." Harvesting of fruit
was regulated so that the tree would not be stunted when it was
young. In addition, fruit trees were not to be cut down as long as
they were bearing, not even as acts of war (Deut. 20:19). Planting
vines (grapes) is mentioned in Isaiah 17:10, and Jesus used the
planting of a fig tree as the basis of a parable (Luke 13:6–9).

Although the ancient religious texts are not primarily biological sourcebooks, we can get some idea of how thoroughly
ancient people understood plants. The germination process was a
familiar mystery:

> The truth is, a kernel of wheat must be planted in the
> soil. Unless it dies it will be alone—a single seed. But
> its death [by germination and growth] will produce

many new kernels—a plentiful harvest of new lives.
(John 12:24, NLT)

## Nurturing the Crop

A famous poem by Solomon begins as follows:

> [1]There is a time for everything, a season for every ac-
> tivity under heaven. [2]A time to be born and a time to
> die. A time to plant and a time to harvest. (Eccles. 3:1–
> 2, NLT)

The time of planting to the time of harvest is a period of crop
nurture in which water supply, weed and pest control, and plant
nutrition are especially emphasized. It is a time of risk, so religion
also acknowledges humble dependence on God in that season:

> So neither he who plants nor he who waters is anything,
> but only God who gives the growth. (1 Cor. 3:7, ESV)

Protecting the crop is the central concern in the season of nurture. First there are the risks of bad weather. Hail is mentioned
thirty times in the Bible. Drought is mentioned at least nine times
and floods at least six times (see especially Hag. 1:10–11). Irrigation and drainage (see Chapter 5) might give some measure of
management, but even today the risk of hail is reduced only by
crop insurance, which does not restore the lost crop.

Human dependence on agriculture was not announced as a
blessing in the Bible. Instead, it says we will get our food by sweat:

> [17]And to Adam he [God] said, ". . . cursed is the ground
> because of you; in pain you shall eat of it all the days of
> your life; [18]thorns and thistles it shall bring forth for you;
> and you shall eat the plants of the field. [19]By the sweat of
> your face you shall eat bread." (Gen. 3:17–19, ESV)

Based on my experience, probably the cause of most of the sweat of
farming has been for weed control. (I am biased by my experience

of pulling cockleburs from cornfields as a boy.) One can sense the pain caused by weeds in this passage:

> [In defiance and despair, Job said,] [38]"If my land has cried out against me and its furrows have wept together, . . . [40]let thorns grow instead of wheat, and foul weeds instead of barley." (Job 31:38, 40, ESV)

Vigilance and diligence against weeds are further emphasized:

> [30]I walked by the field of a lazy person, the vineyard of one lacking sense. [31]I saw that it was overgrown with thorns. It was covered with weeds, and its walls were broken down. (Prov. 24:30–31, NLT)

The prophet Hosea refers to poisonous or noxious weeds (Hos. 10:4), and Jesus tells at least two parables where weeds illustrate an important point. In the parable of the sower (Matt. 13:1–9, 18–23, also in Mark 4 and Luke 8), weeds represent "the cares of the world and the deceitfulness of riches" (ESV). In the parable of the weeds (Matt. 13:24–30, 36–43), they are "the sons of the evil one." The Bible also recognizes that weed problems can become so bad that an infested field is worthless (Jer. 4:3; Heb. 6:8).

Five methods of ancient weed control are mentioned in the scriptures. The simplest method is laboriously pulling them from the ground by hand (Matt. 13:40). Hoeing or chopping off weeds with a hand tool is noted by the prophet Isaiah (Isa. 5:6). He also mentions burning weed-infested fields as a means of control (Isa. 10:17, 27:4, 33:12). A hot fire would kill some of the weed seeds accumulated on the soil surface and in the stubble with mature or nearly mature plants. Weeds also may be pulled and gathered for burning (Matt. 13:40). Jeremiah advises not to plant among thorns but to plow up fallow ground (Jer. 4:3). Thus crop rotation with fallow also is implied to help control weeds. The fifth method is tillage or cultivating the soil. When Adam was sent from Eden, he went out to till the soil (Gen. 3:23). Tillage is a process used to prepare the soil for receiving the seed of a sown crop, but many methods of tillage also tear up the roots of weeds already growing in the field. Plowing is a common form of tillage, and it is mentioned numerous times in the Bible, beginning in Genesis 45:6. It often is regarded as an essential practice for successful agriculture:

> If you are too lazy to plow in the right season, you will have no food at the harvest. (Prov. 20:4, NLT; see also Prov. 12:11)

Plowing is certainly an ancient practice, and though it was difficult work requiring the help of livestock (see Chapter 7), it was widely used because of its role in controlling weeds.

On the negative side, tillage can increase soil erosion and destroy soil structure and organic matter. The negative aspects of agricultural burning and excessive tillage can be reduced today because of technological discoveries of herbicides as well as better machinery. Even so, now, as in the past, the farmer works on hope:

> . . . the plowman should plow in hope and the thresher thresh in hope of sharing in the crop. (1 Cor. 9:10, ESV)

As the modern concepts of integrated pest management have been developed, all of the organisms that can damage or destroy crops have been grouped together as pests. In everyday language, they are weeds, insects, and diseases. In technical detail, they include many kinds of organisms. The Hebrew Bible lists several:

> [If you are disobedient] [39]You shall carry much seed into the field and shall gather in little, for the *locust* shall consume it. [39]You shall plant vineyards and dress them, but you shall neither drink of the wine nor gather the grapes, for the *worm* shall eat them. [40]You shall have olive trees throughout all your territory, but you shall not anoint yourself with the oil, for your olives shall *drop off.* (Deut. 28:38–40, ESV, emphasis added)

Locusts and worms (caterpillars) clearly represent the insects. The locust plague is mentioned more than thirty times in the Bible. Caterpillars are named in 1 Kings 8:37, 2 Chronicles 6:28, and Isaiah 33:4. The dropping of olives may be due to plant disease. Plant disease is referred to as "pestilence or blight or mildew" in

1 Kings 8:37 (see also 2 Chron. 6:28; Amos 4:9; Hag. 2:17) and as "rots" in Joel:

> The seed [grain] is rotten under their clods. (Joel 1:17a, KJV, emphasis added)

The results of an epidemic of a crop disease were certainly understood:

> Ten acres of vineyard will not produce even six gallons of wine. Ten measures of seed will yield only one measure of grain. (Isa. 5:10, NLT)

All of the methods of modern pest control are not mentioned. For example, biocides (or pesticides) were apparently unknown in biblical times. However, the use of landraces of crops adapted to the local stresses that included pests is clearly mentioned in Ezekiel 17:5. Isaiah 28:25 and 2 Samuel 17:28 indicate that several crops were grown at the same time. As a result, the spread of a pest was slowed, and thus a failure of any one crop would not be a disaster:

> Be sure to stay busy and plant a variety of crops, for you never know which will grow—perhaps they all will. (Eccles. 11:6, NLT)

The fallow period in the sabbatical crop rotation (Lev. 25:2–5) also would reduce pest problems (see Chapter 4). Deliberate selection or breeding also is implied in the Bible, primarily for livestock (Gen. 30:40–43). Today, breeding for resistance to pests is an important strategy for integrated pest management. One more aspect of crop management for pest control is fertilization of soil with nutrients that the crop needs for good health and growth. Fertilization of the soil is mentioned in the Bible, but it is so closely associated with the management of livestock that it will be discussed later.

## Harvesting, Storage, and Celebration

The harvesting process has already been mentioned several times. The earliest mention in the Bible is to Cain's harvest of "the fruit of the ground" in Genesis 4:3. The harvest could be bountiful:

> That year Isaac's crops were tremendous! He harvested
> a hundred times more grain than he planted, for the
> LORD blessed him. (Gen. 26:12, NLT)

The increase in seed from planting to harvesting is called the "seed increase rate." It depended on the soil, weather, and kind of crop. It also was quite variable:

> But some seeds fell on fertile soil and produced a crop
> that was thirty, sixty, and even a hundred times as much
> as had been planted. (Matt. 13:8, NLT)

Harvesting also was used as the metaphor of prudence:

> [6]Take a lesson from the ants, you lazybones. Learn from
> their ways and be wise! [7]Even though they have no
> prince, governor, or ruler to make them work, [8]they
> labor hard all summer, gathering food for the winter.
> (Prov. 6:6–8, NLT)

Removing the useful seed from the rest of the harvested plant also is described in some detail:

> [27]Dill is not threshed with a threshing sledge, nor is a
> cart wheel rolled over cumin, but dill is beaten out
> with a stick, and cumin with a rod. [28]Does one crush
> grain for bread? No, he does not thresh it forever;
> when he drives his cart wheel over it with his horses, he
> does not crush it. (Isaiah 28:27–28, ESV)

A modern mechanized farmer would use a combine instead of a stick, a rod, or a threshing sledge, but the same process of grain separation is accomplished. Following separation, grain must be cleaned. The Bible uses the term *winnow* to describe grain being separated from the unwanted chaff (Ruth 3:2). Only then can the grain be safely stored in granaries or barns (Ps. 144:13; Matt. 3:12; Luke 3:17). Grain storage for time of want, even for several years in advance, is described in the old story about Joseph in Egypt (Gen. 41).

After all of the work and worry, a good harvest was a reason to rejoice. Harvest celebrations were woven into the religious calendar of the Bible. Ancient Israel had three. "Firstfruits" (*Yom ha-Bikkurim*, Lev. 23:9–14) anticipated the barley harvest and occurred near the time of Passover (*Pesach)* in the spring. "Weeks" or Pentecost (*Shavu'ot*) occurred fifty days later and celebrated the beginning of the wheat harvest (Lev. 23:15–22). At the end of the agricultural harvest season, there was the feast of "Shelters" (*Sukkot,* also known as "Booths" or "Tabernacles," Lev. 23:33–44), similar in purpose to the American Thanksgiving. Although we usually think of animal sacrifices, grain, oil, and wine were part of the ancient offerings too (Lev. 1-4, 23:13, 18). Bread and wine were essential foods in the celebration festivals carried over into Christianity (Luke 22:19–20). Jesus even said, "I am the bread of life" (John 6:35), comparing the spiritual nourishment he provides to the essential physical nourishment represented by bread. Bread is mentioned 302 times in the Bible (ESV). Wine is mentioned 216 times and oil 197 times. Food is obviously important as a symbol for religious values, probably across all religions. A summary passage about the celebration of food reads as follows:

> And you shall eat and be full, and you shall bless the
> Lord your God for the good land he has given you.
> (Deut. 8:10, ESV)

## The Essentials

An analysis of the biblical data on crops shows that useful crop species are limited in number by adaptation to the local environments where they are to be grown. Anticipating the next chapter, livestock also must be locally adapted. It is true that animals can move around so that they can change their environment in a way that plants cannot, but all of the living organisms that are partners or helpers in agriculture have this requirement of adaptation. It also is interesting that not all of the plants and animals that were locally adapted ended up being useful for agriculture. Some were useful. Most of those were selected and domesticated by humans. Others became pests. All, including the pests, serve some useful ecological function, but from the perspective of the farmer, most local organisms are neither agricultural species nor pests. Nevertheless, the farmer needs a variety of species to help him grow food:

- There must be locally adapted species that can be managed for agriculture.

One might argue that perhaps the statement should apply only to crops and livestock, but some insects, such as honeybees, and some beneficial microorganisms, such as rhizobia (that enhance soil fertility), fit under this principle too.

Under the heading of nurturing crops we considered weeds and pests in general. It might be argued that weeds are so overwhelmingly dominant as a problem that they warrant a separate line: There must be some effective means of weed control. But to shorten the list, I have lumped together all pests:

- There must be effective methods of pest control.

This needs to be understood as applying to all kinds of pests (e.g., the common list of weeds, insects, and diseases) in the process of protecting all kinds of agricultural organisms. Crops, livestock, and whatever other kinds of creatures prove useful or even essential to the successful production of food and other agricultural products need to be protected. The methodology of pest control has developed far beyond the simple procedures described in the Bible. Today's methods often are placed in the categories of natural control (letting nature work for us), biological control (adding biological control agents), cultural practices (e.g., burning and plowing, mentioned earlier), biocides (chemical treatments that kill the pest), and resistance breeding (including several kinds of genetic engineering). Some of these methods are controversial in light of concerns about long-term effectiveness and safety, but both ancient and future farmers must control pests or lose their harvests.

The final point is related to the fact that in most climatic zones where humans live, food cannot be grown all year long. Thus the ancient writings also describe the harvest and storage of food crops. This is more specific to crops than the two previous points because an animal often can be harvested near the time it will be eaten. But not so for plants. The harvest season is short, but we need to eat every day.

- There must be appropriate methods for harvesting and storing crops.

Modern methods often are very different than the ancient methods noted earlier, but the principle is nevertheless very clear. Today we use machines to harvest, clean, and transport our food as never imagined by the biblical authors. We also apply long-distance transport and preservation methods not available in times past. What is clear is that harvest and storage, with their component assumptions about transportation, are two of the focal points about how to do agriculture.

## Conclusions

Food from plants supplies most of the nutrients for most groups of human beings. This is even more true when we realize that the feed for all animals also begins with plants. The underlying fact is that photosynthesis in green plants is the only practical way we know to make food. The dependence of human life on food has layers of meaning. For us to live, something else must die, whether plant or animal. This foundational biological relationship has spiritual significance for many people. Two familiar examples are the Passover meal for the Jew and the Eucharist or Holy Communion for the Christian. Even secular and materialist worldviews must acknowledge that it takes life to sustain life. For us to live, something else must die. Starting with our daily bread as the figurative designation for all of our food, agriculture, and especially crop plants, provides the bulk of our diet. An examination of the accounts and metaphors of the Bible related to crop plants has highlighted these essentials for agriculture:

- There must be locally adapted species that can be managed for agriculture.

- There must be effective methods of pest control.

- There must be appropriate methods for harvesting and storing crops.

Only the last factor is specific to crops. The first two apply to all kinds of organisms useful in agriculture.

# Chapter Seven

## Livestock and Agriculture

### Cowboy Days

Of all the facets of agriculture, I have always strongly identified with livestock and their care. In fact, when I introduce myself, I sometimes say that I am a displaced cowboy. If I am dressed as an "eastern" professor with coat and tie, then my audience may have trouble imagining me with cowboy boots and a wide-brimmed hat. But the statement is true enough. It is true in terms of indelible memories that are as much in my heart as in my mind, memories that recall an array of pictures ranging from raw fury of rangeland weather to beauty and hope represented in scenes from the cycle of life in nature. It also is true in terms of experiences that come from living close to cows and horses in the sparsely populated "West." My life as a cowboy was shortened by other opportunities, but it was long enough and profound enough to have affected my basic identity and tint the glasses of my worldview. My readers should be aware of that.

It is not that I have a particular fondness for horses and cows. They are too much associated with memories of very hard work. I remember going out to deliver hay to our cow herd following blizzards that may have lasted some days. The cows lived outdoors sheltered by windbreaks, which are groves of trees we planted to protect against the prevailing winter winds. The cows would have snow on their backs and snow cakes on their faces where their breath melted the snow, only to have it refreeze in the cold wind. Often after a blizzard, the sky would clear and the temperature would drop to bitter levels. The coldest I remember was −27 degrees Fahrenheit, but my father and brothers saw −40 after I went to college. Today my brother transports and delivers hay from inside a heated tractor cab with machines to place the hay where the cows can eat it. But on that −27 day I remember, we were compelled by the urgency of very hungry cows to brave the elements with only the protection of the very best winter clothes we could find. The clothes were adequate except for the hands. We had to pitch the hay by hand with a fork. To protect our hands, we wore two pairs of mittens, a wool liner inside of a large leather mitten we called a "wood chopper." When it was colder than −15, nothing could keep our hands warm for very long when they were wrapped around the cold wooden handle of a pitchfork for hay. At −27, my fingers would start to go numb in about ten minutes, and there were hours of work to do. To prevent frostbite, we would grab the engine exhaust pipe of the tractor with our mittened hands and hold them there until the leather would smoke with the heat. The smell of charred leather was a part of the job. Then it was back to the hay fork to feed some more cows. It was terrible, especially the pain of warming hands when we were able to come back to the shelter of our heated house. Holding our hands in cold water would help reduce the pain. But our greatest fear was that such weather would come late in winter when the baby calves were being born.

There were days in the spring that compensated for the stress of winter. Between the snowmelt and the greening of grass in the ranges, the cows and their babies were kept close to home. One of the pleasant pictures from my memories is of a group of twelve to twenty baby calves sleeping together with one or two mother cows standing guard. The other mothers would be away for feed or water. The calves would wake and begin a game of follow the leader, running at their top speed with tails in the air and sideways leaps to confuse the followers. The warm spring breezes, the first flowers, and the parade of migrating birds made those days very special.

I never was a rodeo cowboy, exercising special ranch skills in mounted competition, but I well remember the day I "won my spurs," so to speak. Strangely enough, I was on foot at the time. I might have been twelve or thirteen years old, rapidly approaching my adult size. We had used the horses to bring the cow herd into the home corral for sorting into three separate groups for better management. The horses were tied to the fence after the cows were driven into the corral. The process involved letting the cows come down a sorting alley one at a time. There were two side gates. My father was at the first. If a cow was close to giving birth, say in the next month, he would swing his gate across the alley so that the cow would run into a side yard with the rest of that group. If a cow was pregnant but not going to give birth in the next month, I was supposed to open my gate across the alley so that the cow would run into a second side yard with a group of similar animals. If the cow already had a calf, then she was allowed to run the length of the alley to rejoin the calves, which were still outside of the sorting corral.

It was difficult for me to identify the condition of a cow as she came running toward me down the alley. I relied on Dad to give me instructions to let a cow by or sort her into the side yard. I had made several mistakes, and my father was upset with me. Still, I had the less-important position, because it was critical to sort off the cows that were close to giving birth so they could be closely watched and given assistance if needed. My mistakes did not jeopardize a birthing.

The early spring weather had been wet, first with snowmelt and then with rain. The corral was about six inches deep in mud, and we wore work shoes under four-buckle rubber overshoes. We could not be very quick on our feet as a result. Here came a very big cow. The reason was that she was close to calving. My father had let her past his gate by mistake. He called out for me to stop the cow, and I saw that she did not belong in my side yard. With little time to think, I jumped in the middle of the alley, shouting and waving my arms. The cow came on at full speed, lowered her head, and knocked me flat in the mud. Fortunately, she did not step on me, but she exited the corral into the wrong group. My first response was anger, and I tried to run out in front of her again, but she easily escaped because of my heavy boots and the mud.

My father was stunned, I guess because he realized that I could have been seriously injured. We finished the sorting and took the horses out later in the day to recover the one errant cow. That night, bruised and sore, I might have wondered for the first time if it might be possible to do agriculture without cows. That was not a real possibility in the immediate circumstances, and thankfully, my father treated me with a new level of respect whenever we did ranch work together after that. And there was increasing work as I grew older. In addition to working cattle and riding horses, my father taught me to plant trees and to judge and prescribe good management practices for the rangeland. He was a keen student of nature, and in many ways a pioneer in the practices of agricultural sustainability.

One of the compelling propositions of agricultural sustainability is that the best systems will mimic the proven designs of wild nature.[1] Sir Albert Howard,[2] who studied agriculture in India in the early 1900s, is sometimes recognized as the founder of the modern organic farming movement. He thought deeply about natural ecosystems and is attributed with saying, "Mother nature never farms without animals." In wild nature, animals not only provide food for other creatures, their presence diversifies the environment, helping create additional niches for other organisms, and their waste products facilitate the recycling and transfer of soil nutrients through the food webs of their home ecosystems. One example of the transfer of nutrients is the natural consumption of fish by eagles that thus move nutrients accumulating in aquatic systems back to terrestrial nesting sites.

## Are Livestock Really Good for Us and the Environment?

When applied to agriculture, the ecological roles of livestock encompass the same functions as animals in the wild: food (meat, eggs, dairy), diversification, and nutrient cycling and transfers. Before looking at scriptural insights about livestock in agriculture, a brief overview of some of the current discussions about these topics is in order. The first point involves the impact of livestock on the supply of food for humans. Some have reasoned that livestock must decrease the total food supply because they require feed grains that could otherwise be used by humans. If humans do not eat those particular grains, then the land used to grow them could be used for alternative crops that humans could eat. The fundamental ecological and biochemical fact behind this argument is that livestock are inefficient converters of the food they eat. Only 5 to 30 percent of the food energy eaten by livestock ends up as meat, eggs, or dairy suitable for the human diet.[3] Thus to the extent that farm animals eat grains and otherwise use land suitable for the production of food for humans, they do have a negative impact on the total food supply.

However, that is not a complete picture. As stated in Chapter 4, only about 11 percent of the earth's total land area is suited for the sustained production of food crops for humans. Much of the earth's land is too steep, too rocky, too wet, or too dry to be useful in the direct production of food crops. But some of this other land can be used as pasture or range, given careful management and stewardship. In addition, a significant proportion of the "good land" needs to be in crop rotations with seasons of legumes and grasses that regenerate the soil. Although these crops can be used as "green manures" that are plowed back into the soil, they also can be used as forages to feed livestock. Finally, our food production systems produce "by-products" that are not suitable as human food but that can be used as livestock feed. Sugar beet pulp (from sugar production), citrus pulp (from orange juice production), distillers' grains (from alcohol production—for drinking or fuel), and cornstalks (from corn grain production) can all be used as parts of a livestock ration. The result is that livestock can be used in agriculture to increase the food supply.[4] Modern grain-based livestock production systems have environmental and nutritional disadvantages, but there are ways of producing meat, dairy, and eggs that do not have those disadvantages.

Another related discussion is the nutritional consequences of consuming animal-based foods. There is growing evidence that the products from animals with a "natural diet" high in forages (as opposed to grains) are in fact beneficial for human health.[5] From the perspective of food quantity and quality, much of our animal-based agriculture needs to be redesigned, but there are sound nutritional and agricultural principles that favor the use of animals in our food production systems.

The second point relates to the role of livestock in diversifying the agricultural landscape. Growing several kinds of crops and changing the fields in which they are grown is the basis of crop rotation. Crop rotation is one of the key practices in designing pest control strategies that do not rely on biocides (see Chapter 6). Grazing livestock are able to utilize some plants that would otherwise be weeds, and the high level of plant diversity common in grazed lands leads to complex insect populations that include natural enemies of many pests. Balancing livestock rations favors the production of various kinds of crops, including grass and legume forages that help restore farmland that has been stressed by intensive cultivation. Some of the best crops for naturally enriching the soil nitrogen supply are also the best crops for feeding livestock. One of the most sustainable agricultural systems known is the mixed cropping system that developed in Europe and Asia, perhaps 7000 years ago. It included wheat, barley, peas, lentils, and cattle and sheep.[6] One of the probable reasons for the long success of this system is the diversity it entails.

Another reason for the success of the mixed cropping system was the nutrient transfer and cycling practices it used before the development of the fertilizer industry. Without commercial fertilizers, every sustainable cropping system still must include a means of replenishing soil nutrients. For flooded rice in Asia and irrigated wheat in Egypt and the Middle East, natural soil erosion from the uplands followed by deposition on the floodplains maintained the soil nutrients for agriculture along the rivers. But for upland areas, another approach was needed. Livestock manures were used to carry soil nutrients to where they were needed, often exporting nutrients accumulating in lowland pastures and meadows back to upland grain fields and stubbles.[7] In its simplest version, livestock were allowed to graze on stubble fields following the grain harvest. When rotated from nutrient-enriched lowland pastures in the day to nutrient-depleted stubbles at night, there was a net transfer of nutrients to the stubbles, because animals pass more waste at night. The process might be enhanced by taking supplemental feed (e.g., hay) to the animals on the stubble fields. In addition, the manure of animals confined for milking or winter feeding could be carried to the grain-producing fields. In temperate climates, livestock require winter feeding, and the collection and spreading of livestock manure from confined livestock is an important feature of many winter feeding systems. In ancient times, the winter feed of hay was harvested by hand with the scythe, often from lowland meadows where soil nutrients were naturally deposited during flooding.[8] One could argue that the most important consequence of the Iron Age was the development of the iron scythe, which allowed the massive expansion of the mixed cropping system based on nutrient management by means of the winter feeding of livestock. The expansion of European culture around the world followed.[9] At the heart of the success of the European agricultural system was the use of livestock to sustain the soil nutrient supplies. The textbook by Peter Cheeke provides a more detailed analysis of the role of livestock in agricultural sustainability and the food system.[10]

## Livestock in the Ancient Writings

The biblical perspective on livestock is in the context of the mixed cropping system that has just been outlined. The Bible has more to say about livestock than any other agricultural topic. Livestock were economically, nutritionally, and religiously important to the ancient peoples who are the spiritual ancestors of modern Jews, Christians, and Muslims. The role of animal sacrifice in worship was established in the Bible before animals were proclaimed to be also good for food (Gen. 4:4 and Gen. 9:2–3). Some people may today be offended by the thought of animal sacrifices, and the practice certainly raises questions about the treatment of animals. But the biblical perspective lends strong support for the humane treatment of livestock. First, God is represented as caring about animals, especially in the account of Noah and the ark:

> But God remembered Noah and all the beasts and all the livestock that were with him in the ark. (Gen. 8:1a, ESV)

Second, proper human behavior is presented as fostering animal welfare:

> The godly are concerned for the welfare of their animals. (Prov. 12:10a, NLT)

Third, the mistreatment of animals is condemned:

> [6]... they crippled oxen just for sport. [7]Cursed be their
> anger, for it is fierce, cursed be their wrath, for it is
> cruel. (Gen. 49:6–7, NLT)

Trapping wild animals to terrify them also is condemned (Hab. 2:17). When humans were commanded to rest every seven days, they were to give their working livestock rest as well (Exod. 23:12; Deut. 5:14). In addition, humans were to care for their livestock even on the days when they otherwise rested (Matt. 12:11; Luke 13:15, 14:5), and a working animal was to be allowed to eat while it was working (Deut. 25:4; 1 Cor. 9:9; 1 Tim. 5:18). Even the livestock that belonged to one's enemies were to be helped if they were overworked (Exod. 23:5).

With this concern for animal welfare, it is not surprising that the slaughter of animals also is covered in the Bible. The perspective is that killing animals is a concern of God. Animal slaughter was not something that could be done by just anyone when and how one pleased. A skill and respect were required for slaughtering animals (Lev. 17:3–4; Deut. 12:15–16, 20–25), and animals "torn by beasts in the field" were not to be eaten by people (Exod. 22:31; Lev. 22:8; Ezek. 4:14).[11] Although the Hebrew Bible law required that firstborn cattle and sheep be offered as a sacrifice to God, they could be eaten in the proper circumstances (Deut. 15:19–20). However, they could not be worked or sheared (Deut. 15:19), and they were to be allowed to nurse for at least seven days after being born (Exod. 22:30).[12] The humane killing of animals is a part of the Jewish (*kosher*) and Islamic (*halal*) traditions. It is an anomaly that religious and ethical concerns are not more prominent in the modern Western meatpacking industry.

## Other Aspects of Animal Welfare

The good treatment of animals extended to other situations as well. Livestock that stray from their owners or caretakers are not only a risk to the property and persons of others (Exod. 22:5), they also are unguarded and thus at risk of being attacked by wild beasts or uncontrolled dogs. There are laws in the Bible for these situations too:

> ¹You shall not see your brother's ox or his sheep going astray and ignore them. You shall take them back to your brother. ²And if he does not live near you and you do not know who he is, you shall bring it home to your house, and it shall stay with you until your brother seeks it. Then you shall restore it to him. (Deut. 22:1–2, ESV)

A related passage makes it clear that this was not just an economic law for the benefit of the community of "brothers" but that it was based on the principles of animal care:

> ⁴If you meet your enemy's ox or his donkey going astray, you shall bring it back to him. ⁵If you see the donkey of one who hates you lying down under its burden, you shall refrain from leaving him with it; you shall rescue it with him. (Exod. 23:4–5, ESV)

It also should be noted in passing that dogs are mentioned thirty-four times in Scripture, mostly in a negative context of semiferal animals. However, Job 30:1 makes reference to dogs that help care for the flock, and Mark 7:28 refers to dogs under the table where children are eating. Thus pets and companion animals are noted in the Bible.

Any person who has worked with animals knows that they can be dangerous. They can even kill other animals or people. Biblical regulations cover the various situations and responsibilities that might arise. The underlying rules are as follows:

> ³⁵If someone's bull injures a neighbor's bull and the injured bull dies, then the two owners must sell the live bull and divide the money between them. Each will also own half of the dead bull. ³⁶But if the bull was known from past experience to gore, yet its owner failed to keep it under control, the money will not be divided. The owner of the living bull must pay in full for

> the dead bull but then gets to keep it. (Exod. 21:
> 35–36, NLT)

> ... Animals that kill people must die. (Gen. 9:5b, NLT;
> see also Exod. 21:28–32)

The humane treatment of animals is represented even in this. A dangerous animal must be so confined that it cannot live a normal life. It will be miserable and increasingly dangerous. Ending its life is the best thing to do.

The animals that are most dangerous are mothers protecting their young and males guarding their mates. The Bible does not give much information about the management of these natural processes. Certainly Jacob tried to control the breeding of the livestock he managed (Gen. 30:39, 31:10–12), but his methods do not seem scientifically sound. (Nevertheless, God blessed his methods.) We also know that shepherds watched their flocks to guard them from danger (1 Sam. 17:34; Luke 2:8), and that Jesus compared himself to a shepherd (John 10:11, 14). Ancient shepherds probably managed breeding and birthing just like their modern counterparts. The Bible does forbid unnatural breeding, which might amplify animal stress.

> ... Do not breed your cattle with other kinds of animals. (Lev. 19:19, NLT)

However, it also is clear that this does not apply to all kinds of crossing, because mules (the cross of a donkey and a horse) are mentioned numerous times in the Bible, starting in 2 Sam. 13:29, without a negative connotation.

## Economic and Religious Factors

The domestication of livestock implies that they have been selected through generations to become more docile and dependent on humans for feed and protection. Hiebert[13] reasons that when God brought the animals to Adam to name and search for a helper (Gen. 2:18–20), Adam, the first human, actually did select the livestock species as his helpers from all of the kinds of animals. All animals were approved for meat (Gen. 9:3; Acts 10:11–16; 1 Tim. 4:4), but only domesticated animals provide milk and power for work. Milk is mentioned forty-seven times in the English Standard Version of the Bible, and seventeen times in reference to a land "flowing with milk and honey," the place of great blessing. Curds (similar to yogurt or cottage cheese) are mentioned eight times, and cheese is mentioned twice. References to horses, donkeys, camels, and mules carrying people or baggage are very numerous, and oxen plowing is mentioned at least three times (1 Kings 19:19; Job 1:14; Amos 6:12). In English usage, oxen are castrated male cattle, emasculated to control the aggressive behaviorial characteristic of bulls. However, the Hebrew words translated "oxen" refer to both male and female animals, and the rabbinic interpretation is that Hebrew livestock were not castrated. The relevant verse can be interpreted two ways:

> Any animal that has its testicles bruised or crushed or torn or cut you shall not offer to the LORD; you shall not do it within your land. (Lev. 22:24, ESV)

The current Jewish interpretation is that all castration is forbidden.[14] Alternatively, the last phrase could be regarded as a simple redundancy forbidding only the offering of castrated animals. Oxen (or the singular ox) are mentioned 137 times in the Bible. Bulls (noncastrated males) are mentioned 114 times, often in the context of an animal sacrifice.

As sources of food and preindustrial power and transportation, livestock were of course economically important in the past as they are today. Livestock are still the financial savings accounts for cultures without well-developed banking systems. They also are the main source of supplemental energy for human work in many cultures, including the Amish in America. Instead of fossil fuels for tractors, the Amish energy system depends on recent solar radiation to grow hay and oats to feed their horses.[15]

Thus there was and is a human-centered motivation for good animal care. The biblical design of villages, towns, and cities included sheepfolds within the confines of the settlement (see especially 2 Chron. 32:27–29) and nearby grazing lands (Lev. 25:34; Num. 32:24, 36, 35:2–4) so that they could provide the shelter and feed needed by their livestock.

The economic value of livestock also is associated with a whole set of biblical rules governing ownership and loss of animals. The general principles of ownership will be discussed in a later chapter, but the central concepts with regard to livestock are summarized as follows:

> [18]Anyone who kills another person's animal must pay it back in full—a live animal for the animal that was killed.... [21]Whoever kills an animal must make full restitution. (Lev. 24:18, 21, NLT)

There were rules covering the accidental death or injury of livestock (Exod. 21:33–34), for settling disputed ownership (Exod. 22:9; Deut. 25:1), for compensating for damage caused by livestock under one's care (Exod. 22:12–13), and for lending or renting livestock (Exod. 22:14–15). Perhaps, livestock were increasingly valued with the passage of time. The monetary fine for stealing livestock set in the book of Exodus, supposedly written at the time of Moses, was five times the value of a beast of burden and four times the value of other animals (Exod. 22:1, 4). In the book of Proverbs, supposedly written several hundred years later at the time of King Solomon, the general fine for stealing was set at seven times the value of the stolen property (Prov. 6:31).

Although layered in spiritual meaning, the economic value of livestock was an important contributing factor toward their role in the religious life of the ancient communities. Sacrificing one's livestock was a real sacrifice. As a Christian colleague once commented to me, a burnt offering of a bull in Bible times would be economically similar to sacrificing one's car today.

To sum up, the biblical perspective is that all animals are valued by God and subject to human stewardship. In addition, domesticated animals have special value but are in a dependent relationship with humans. They thus should be subject to special care. In summarizing the biblical standard for the treatment of animals, my colleague Robert De Haan argued that all animals are to be allowed to live so that their God-given characteristics are recognized and celebrated. "They were created to form social relationships, to mate, to nest, and to raise their young . . . [thus] giving glory and praise to God the Creator."[16]

> Let the sea and everything in it shout his praise! Let
> the earth and *all living things* join in. (Psalm 98:7, NLT,
> emphasis added)

Taking a secular approach, Michael Pollan reached the same conclusion: that a domesticated animal exhibits "animal happiness" when allowed "to express its creaturely character."[17]

## The Essentials

Two agricultural essentials emerge from the biblical perspective and a third can be recognized with a bit of reflective thinking. The first is that livestock are an important part of agriculture, and the second is that livestock are to be well cared for. Both points need further explanatory qualifications.

The Bible might be interpreted to imply that livestock are a part of all agriculture, but some examples show that this might not be true. Modern cropping systems without livestock can be designed with power and fertilizers extracted from fossil fuels. Future solar and atomic power might make this even more sustainable. Dietary vegans (consuming no animal products) and animal rights advocates might support such systems, and efforts to maximize food production for very high human populations might dictate minimizing livestock contributions. This brings us back to the fact that the Bible is not a textbook on agriculture, and even that its examples are framed in a specific, ancient context. In the case of livestock, this means more than just changing the breeds and species to fit other conditions. There may be sustainable agricultural systems that do not include livestock. Thus the essential point is not that livestock are required. A more precise statement is this:

- Livestock can make important ecological, nutritional, economic, and cultural contributions to our agricultural systems.

I will say more about the cultural contributions in a later chapter, but the role of livestock in religion has been outlined earlier.

The second and related essential point is a clear emphasis of the biblical perspective:

- Animal welfare is a central concern of agriculture, and all livestock should be well cared for.

The biblical perspective, however, does give room to manage livestock by controlled breeding and crossing, for example, to produce new breeds and hybrids as in the case of mules. The overall view is that all animals as living, sentient creatures are to be respected and allowed to live as naturally as possible. In the case of livestock, a symbiotic relationship with humans has developed where animals get protection, feed, and care, and humans get food, power, and animal companionship. Livestock are thus very much our partners in the endeavor of agriculture.

Finally, there is the more subtle point that prior to the modern industrial era with energy supplied by fossil fuels, livestock were main sources of supplemental power for agriculture and commerce. Our history books tell us that story, and the Bible confirms its long history. There is now uncertainty about future sources of energy, but many farming systems around the world, including Amish farming in the United States and Canada, still get their horsepower from horses or oxen, livestock that in turn get their energy from feed grown with recent solar radiation. Thus there is another essential of agriculture seen in considering livestock:

- Animal traction is a model of a sustainable means of supplying the required energy for agriculture.

If necessary, a partial or complete restoration of animal energy for agriculture could be made, but the transfer of land and resources to care for the increased number of animals that would be required would reduce the number of people agriculture could feed. In any case, the animal model for agricultural energy is an example of using crops and forages instead of fossil fuels as our main source of energy. It is a renewable energy system.

## Conclusions

Livestock are important in many agricultural systems. They contribute food, supplemental power for farming and transportation, and ecological services in facilitating the nutrient cycles through management of manure distribution to agricultural lands. They serve not only as partners in the agricultural endeavor but also as welcome companions for many people. The biblical overview of animals in agriculture emphasizes the following essentials of agricultural sustainability:

- Livestock can make important ecological, nutritional, economic, and cultural contributions to our agricultural systems.

- Animal welfare is a central concern of agriculture, and all livestock should be well cared for.

- Animal traction is a model of a sustainable means of supplying the required energy for agriculture.

Natural ecosystems always include animals. Although agriculture might be made to function without livestock, to do so would be to deviate from the natural model of agriculture.

# Chapter Eight

# Farming Systems and the Practices of Agriculture

## A Good Mentor

It is a great advantage to have a good mentor, and I am fortunate because I have had several good mentors. One of the times in my life when I really needed that kind of help was when I accepted my position at Cornell University as a starting assistant professor. Cornell is located in the Finger Lakes region of New York State. My wife and I moved there following the completion of my graduate studies in California. It was a totally new environment for us with ill-defined job expectations. My new department was looking for the unconventional, and they thought I was qualified. It was the only good job offer I had in a sparse job market, or I probably would have looked for something else I could have managed with more confidence. Successfully finding my niche and satisfying the requirements for tenure took some help. One of my best helpers and mentors at the time was Professor Bob Lucey, who passed away in 2004.

My job required that I do research and offer classes that would help the present and future farmers of New York State. Professor Lucey had similar general requirements for his job, though his focus was working directly with farmers in "the North Country." To help me acclimate, he made it possible for me sometimes to travel with him as he visited his research sites and cooperating farmers up north. We would travel together for two or three days at a time. Just being with him on the long drives gave him time to explain the regional agriculture to me and to answer my many questions. More importantly, I was exposed to his philosophy of designing research and interacting with people. His research goal was to make a difference in the total quality of life for the farmers in the region. He started with soils that were too wet and put together a team of cooperators to technically and practically solve the problem. Then he advanced through the next issues until his program pretty much addressed the whole farming system, including its social and economic components.

I gained the best understanding of his approach and of his message for me by going with him on farm visits. He really knew and cared about the farmers and their entire families. Bob's car was recognized and welcomed by many farmers whether he was expected or not. He greeted each contact by name or made a warm and sincere introduction. He would ask about crops and cows and plans. He would ask about the farmer's family by name. Sometimes he would ask about a prize cow by name. It was amazing to watch him use his personal skills to gain trust and uncover real needs. And it was not that he was just a "smooth operator" who wanted to make you feel good while you were together. He was one of the few true "systems thinkers" I have known. He was thinking about soils and drainage and fertilizers all at once. At the same time, his vision encompassed crops and cows and markets and people and the social interactions of people at fairs and churches and schools. He had the kind of mind that saw the big picture and the important details together, and he had the clear goal in mind of making farm life better. In a word, he was holistic. He could think about whole farms, and he could get teams together to work on the problems of whole farms. Though he and I eventually worked in different ways, he inspired and mentored me to think about systems and keep the whole farm in view. Bob had a great career and was widely recognized for leadership in improving North Country agriculture. He eventually was recognized with an endowed chair at Cornell University, the highest honor the university can give. It was the first such recognition ever given to a professor with primary responsibilities in agricultural extension and public service.

In this chapter, my goal is to introduce the broad view I learned from Bob Lucey. It includes the physical processes of farming as they apply to whole farms. The topics addressed here have in common the inclusion of several dimensions and components of agriculture as they are interrelated, both in terms of biology and management provided by the farmer. Stated simply, we need to think about soil, crops, and livestock all at the same time. We need to think about the practices of agriculture that interlink and coordinate all of the parts and processes of a whole farm. In short, we need to think about the farming system.

## Systems Thinking

Systems thinking involves both analysis and synthesis, finding out what are the parts and how they are put together. Biologists do this in terms of physiology (a reductionistic approach) and ecology (a holistic or an integrative approach). Systems scientists present this idea in terms of the hierarchy of biological organization: cells organized into organs and tissues, tissues into organisms, organisms into populations and communities, communities into ecosystems, and so on. Each level in the hierarchy partly "explains" some higher or more inclusive level and is itself "explained by" lower or less inclusive levels. However, such explanations are incomplete, because there are unpredictable characteristics called "emergent properties" that come from the interactions within an integrated whole. As proposed by the ancient Greek philosophers, the whole is more than the sum of its parts. This truth is most easily recognized when it comes to human actions and interactions. Thinking, and especially the details of the thoughts themselves, is not revealed in a biological and physical analysis of the human brain. The exact effect of education, or religion, or enlightened self interest on any kind of human behavior, including environmental care and farming practices, is not clear from the analysis of each aspect of human nature and culture, or even from the dissection of all of the factors taken together.

What can be done with systems thinking is to describe and summarize, often using mathematical relationships and statistical probabilities to approximate the behavior of real systems. For the subject at hand, agriculture works or does not work mostly because of how its parts are put together. Systems thinking is what we need to understand the mechanisms of success and failure in agricultural systems.

There is a large body of knowledge related to the study of agricultural systems. The academic titles on the technical books about farming systems may hide the fact that each farmer is a systems thinker; he or she must visualize what he or she wants to do and then synchronize all of the parts and processes for getting there. As a minimal example, having livestock also involves providing feed for the livestock and disposing of the wastes that livestock produce, usually by returning them to the soil. Producing feed involves sustaining soil health and productivity over time spans as short as a few years to as long as multiple generations. The same is true for directly producing food for one's family or for the market. Sustaining soil health involves the management of life cycles and nutrient cycles that loop through other levels of system organization.

The concept of systems thinking as the core discipline of management science is relatively new, but rough examples can be found in the ancient writings, especially as they relate to the practices of farming. In addition to a general overview of systems thinking from the perspective of the Bible, two biblical topics provide rudimentary examples—soil tillage and nutrient management. Neither is fully comprehensive or holistic in terms of the whole of agriculture, but both involve aspects of the management of soil, crops, and livestock taken together. In addition, both are relevant in terms of the challenges to develop agricultural systems that are more sustainable than common current practices. I will give a brief summary of the current concerns before examining the biblical perspectives.


## Soil Tillage

To till the soil means to prepare it or maintain it for producing crops. Some mechanical mixing or manipulation of the soil is involved, as in plowing, fertilizer injection, and mechanical weed control with cultivators. The scale of farming (size of fields) is dependent on the energy supply to do tillage. Energy from mechanical tractors allowed the scale to increase compared to traction energy from livestock. Likewise, livestock energy allowed larger fields to be worked than energy from human labor alone.

Conventional plowing is the complete inversion of the soil surface, usually to a depth of six to eight inches (fifteen to twenty cm), and it helps control weeds and incorporate surface nutrients and residues into the soil. Pests in the residues also may be controlled by plowing. However, conventional plowing also exposes nearly bare soil to water and wind erosion, and significant agricultural research in the past seventy years has been aimed at developing alternatives that provide more protection from erosion. As alternative tillage methods have been more widely used, it has become clear that not only do they reduce erosion and promote water penetration into the soil, they also allow the deep percolation of agricultural chemicals, some of them undesirable nutrients and pesticides, through undisturbed root and animal tunnels in the soil. Since plowing helps control many pests, reduced plowing has led to the use of more pesticides. In turn, these pesticides have been able to reach groundwater supplies more directly. The challenge is to find sustainable means of protecting the water, the soil, and the crops, all at the same time. At least a partial solution to this problem is possible with the use of soil mulches, which can be provided by cover crops and "green manures" grown between seasons and in fallow periods. A mulch on the soil surface helps control weeds, protects from soil erosion, and actually holds some nutrients so they cannot be carried by water to surface and subsoil water supplies.

Soil tillage also is used to form a good seedbed for planting a crop. To germinate in the soil, a seed needs to be planted at a proper depth (based on its size), where it has enough biochemical (food) energy in the seed to penetrate its roots to soil water and emerge its shoot to sunlight. Firm contact with the soil facilitates penetration and emergence, so the soil needs to be manipulated so as not to be too cloddy, or fluffy, or pulverized to dust that will crust when wetted. In conservation tillage, which leaves residues on the surface to protect from erosion, only a narrow strip of soil may be tilled to form a seedbed. Many modern drills (machines that plant crop seed) place the seed in rows in the soil at a roughly controlled depth with fertilizer nutrients and pesticides "banded" in the soil near the planted seed. The preparation of the seed bed and the actual planting of the seed may involve multiple trips across the field, or it may be accomplished in a single operation, depending on the tillage methods and machines in use. The use of tools and supplemental energy (from livestock or machines) has made farmwork less strenuous and more productive. At the same time, big machines associated with controlled tillage and planting have favored big farms and less farm employment.

## Nutrient Management

Natural ecosystems run on the flow of energy, supplied in the first step by sunshine, and on the circulation of matter through the nutrient and water cycles. Agriculture diverts some of the energy and nutrient materials to food and other useful products. The energy supply from the sun is not diminished by agriculture, but the export of nutrients in food will eventually cause productivity to decline and reach levels too low for practical harvesting, unless the nutrients are replaced. Nutrient replacement or management is thus a part of all sustainable agricultural systems. The discoveries of chemistry and the Industrial Revolution eventually led to the science and practice of supplying crop nutrients in fertilizers that were manufactured and transported to the farm, often from great distances. Even as the use of manufactured fertilizers was increasing at a rapid rate, cautious leaders were advocating the more natural practices from traditional farming methods. The approaches that attempted to mimic nature became known as "organic farming," and in general, organic farmers avoided the use of all manufactured chemicals that were not naturally occurring materials. As some natural products were found to be harmful to the environment, they also were banned from organic farming.[1]

The two contrasting approaches and philosophies of nutrient management led to a great agricultural debate. The industrial or manufacturing proponents argued that soil and plant science showed (correctly) that crop nutrients entered plant roots in an inorganic form. Thus organic farmers were not really organic at a critical step in the process. Furthermore, the traditional methods could not supply enough nutrients fast enough to give the yields needed by a world threatened with famine. The "Green Revolution" in the last quarter of the twentieth century developed from genetic changes in the main crops that allowed them to utilize high rates of fertilization. It can be argued that the Green Revolution saved more human lives from starvation than any other human endeavor, ever.

At the beginning of the debate, the organic side mostly had to argue that precaution was warranted when the consequences of new practices were unknown. With the passage of time, negative effects on soil biology and the pollution of ground and surface waters by excessive nutrients have strengthened their position. In addition, the argument that organic agriculture has low productivity is weakened by the fact that organic agriculture has never benefited from the research and development that supported industrial-based agriculture. Some recent scientific studies reflect favorably on the productivity of organic methods,[2, 3] and the influence of organic farming has spread around the world.[4] At the beginning of the twenty-first century, there is the real concern that the industrial model for agriculture is good for industry but bad for rural communities and ecosystems. More will be said about this in the next chapter, but it should be clear that these topics have multiple dimensions.

Yet another dimension of the argument about nutrient management is related to the role of natural livestock manures. Animal manures facilitate the natural cycling of nutrients, and most are good crop fertilizers. Yet as livestock are concentrated for more efficient meat, egg, and dairy production, their feed must be imported so that there is also a concentration of nutrients in manure that exceeds the local environment's capacity to hold and utilize those nutrients. Nitrogen escapes as ammonia to the air or leaks as nitrate to groundwater, phosphorus is carried by soil erosion into surface waters, and potassium accumulates in the soil. Very high soil potassium levels can result in forages that are actually toxic to cattle and sheep.[5] Fortunately, the environmental injury so far is not overwhelming, but the direction of change is negative. Many governments are already forced to spend

Shelton State Libraries
Shelton State Community College

significant amounts of money to encourage or force farmers to protect the environment from the nutrients they import to their farms as fertilizers and livestock feeds.

With the industrial model of farming, agriculture has become a very complicated and technical endeavor. How are farmers going to learn to farm? How should they learn to farm? The modern answer is to turn more and more to formal education as a complement to the traditional experience of learning how to farm as it is interpreted and mentored across the generations. Masses of information and data about how to farm are becoming available. Tillage machines and harvesting combines can be equipped to provide "site specific management" for fertilization, pest control, and irrigation. The World Wide Web has been added to conventional print media to supply the interested farmer with knowledge and advertising about the latest available options. But world wide, there is a spectrum of farmers. Some are still illiterate peasants, and the ancient wisdom of the Bible offers additional insight on what is needed.

## The Wisdom for Farming

Dick and Sharon Thompson are Iowa farmers widely recognized for their leadership in advancing the methods of agricultural sustainability. They have boldly acknowledged that God is the one who is teaching them how to farm.[6] In fact the Bible asserts that farming takes both wisdom and revelation and that God is the primary teacher.

> [26]The farmer knows just what to do, for God has given him understanding. . . . [29]The LORD Almighty is a wonderful teacher, and he gives the farmer great wisdom. (Isa. 28:26,29, NLT)

The details of similar passages outline some of the practices of ancient Near Eastern farming (Ps. 107:37; Prov. 27:25–27; Isa. 28:25–29), mysterious to the writer of those parts of the Bible but portions of the local wisdom needed for farming. The annual pattern of sowing, growing, and harvesting is a common theme. The Bible further shows that fear of God (meaning respect and honor that lead to obedience) is the foundation of all wisdom:

> The fear of the Lord is the beginning of wisdom, and the knowledge of the Holy One is insight. (Prov. 9:10, ESV; see also Ps. 111:10; and Prov. 1:7)

(From a biblical perspective, this is the basis for a "wise-use" policy for natural resources.) The Genesis account of the great seven-year famine further attributes Joseph's understanding of how to prepare for the crisis to wisdom that came from God (Gen. 41:16, 39). Of course, much of this wisdom for farming was localized and passed on from generation to generation as farming parents taught farming children what they needed to know and understand. Such wisdom is based on the concept of biblical stewardship outlined in Chapter 2. It is significant that the Bible stresses wisdom (knowing what to do) as opposed to knowledge (just knowing facts). Knowledge also is highly regarded in the scriptures, but wisdom is emphasized. The biblical picture of the farmer is not just of one who knows the methods and the seasons (facts) but of one who is in a prayerful, attentive relationship with the Creator, seeking wisdom for all the questions of farming. This is comprehensive and links the care and management of soil, crops, and livestock considered in this chapter. This comprehensive viewpoint is summarized in the New Testament as the Kingdom of God or the Kingdom of Heaven proclaimed by Jesus Christ and his followers. It places greatest emphasis on righteousness (right relationships) and on human values:

> For the kingdom of God is not a matter of eating and drinking but of righteousness and peace and joy in the Holy Spirit. (Rom. 14:17, ESV)

Modern poet and agricultural philosopher Wendell Berry refers to a similar coherent and comprehensive viewpoint that he calls "the Great Economy."[7]

## Soil Tillage in the Bible

The process of plowing is mentioned about fifteen times in the ESV translation. Plowing is a coarse tillage, and the resulting rough surface of the soil may be smoothed by harrowing, which is a finer tillage procedure mentioned three times in the Bible.

The prophet Isaiah asks rhetorical questions that reflect the importance of tillage in biblical times:

> Does a farmer always plow and never sow? Is he forever cultivating the soil and never planting it? (Isa. 28:24, NLT; see also 1 Cor. 9:10)

The need to plow on time or to reduce the amount of plowing (reduced tillage) is implied here. The prophet Jeremiah used an analogy of plowing that shows it was used on hard ground:

> This is what the LORD says to the people of Judah and Jerusalem: "Plow up the hard ground of your hearts! Do not waste your good seed among thorns." (Jer. 4:3, NLT)

Tilling the soil also is linked by Scripture to prosperity:

> Whoever works [or tills] his land will have plenty of bread. (Prov. 12:11, ESV)

> If you are too lazy to plow in the right season, you will have no food at the harvest. (Prov. 20:4, NLT)

Following plowing and harrowing, the seed of cereals was broadcast onto the soil surface by the hand of the sower (the farmworker sowing seed). Jesus told a famous parable of the sower who scattered seed among the rocks, weeds, and along the path as well as on good ground (Matt. 13:1–9; Mark 4:1–8; Luke 8:4–8). For the seed to germinate and establish an adequate root system, there must be good contact with the soil, and most seeds need to be firmed into the seedbed and slightly covered with soil. MacKay[8] made the appropriate interpretation of Isaiah 7:25 and 32:20:

Cattle and sheep were allowed to tread the seed into the ground on newly sown fields. This practice would firm the soil, if it was not too wet, and give good seed-soil contact. My father used the same practice on our home ranch to restore native meadows on worn-out cropland.

One of the characteristics of soil tillage is that it takes a great deal of energy. It appears that a pair (yoke) of oxen was commonly used to plow in biblical times. Elisha plowed with twelve yoke of oxen (1 Kings 19:19), and Jesus refers to a man who purchased five yoke oxen (Luke 14:19). Job, who was apparently one of the richest farmers of the Bible, owned 1000 yoke of oxen (Job 42:12). Donkeys also could be used for plowing (Deut. 22:10), but they are mentioned more often for their work in carrying loads. It should be noted that traction energy from livestock characterized the old farming systems of Africa, Asia, and Europe. The Americas and the Pacific Islands developed fine agricultural systems with only human labor, but they were also supplanted by the European systems when more energy-intensive options were available.

## Nutrient Management in the Bible

The natural fertility of valleys and wetlands is mentioned early in the Bible:

> Lot took a long look at the *fertile plains* of the Jordan Valley in the direction of Zoar. The whole area was *well watered everywhere*, like the garden of the LORD or the beautiful land of Egypt. (Gen. 13:10a, NLT, emphasis added)

Ezekiel identified fertile soil and abundant waters as the needed factors for luxuriant crop growth (Ezek. 17:5). As mentioned earlier, nutrients that erode from uplands are accumulated in lowlands, thus replenishing what might be removed by cropping. When nutrients are not replenished, the Bible also describes the results:

> They have sown wheat and have reaped thorns; they
> have tired themselves out but profit nothing. They shall
> be ashamed of their harvests. (Jer. 12:13a, ESV)

This passage refers to general mismanagement, but the picture is one of infertility.

The importance of fertilizer for crop production is emphasized in Jesus' parable of the fig tree. The gardener pleads to give the unproductive fig tree one more year, saying he will "give it special attention and plenty of fertilizer" (Luke 13:8, NLT), or that he will "dig around it and put on manure" (ESV). In several places, the Bible says that dead bodies are like manure or dung spread across the ground (2 Kings 9:37; Ps. 83:10; Jer. 9:22, 16:4, 25:33). This implies that livestock manure was being used for fertilizer, and that dead animal bodies were understood to have a similar effect. The dead animals might have first been burned (Exod. 29:14 and elsewhere). Ashes from fires were apparently used as fertilizer (Ps. 147:16). Livestock manure was composted by working in straw (Isa. 25:10), and various wastes were added to the manure compost pile (Luke 14:34–35). The passage in Luke 14 indicates that salt may have been used in fertilizer. This is puzzling, because the ancients certainly understood that salt in large amounts was bad for the soil. Abimelech salted the soil of his enemies (Judg. 9:45), and salty soil was viewed as a curse (Ps. 107:34).

## The Essentials

The concepts of soil tillage and nutrient management are not well developed in the Bible. However, enough information is provided that we can see that systems thinking about farming practices was being developed even in ancient times. These topics that encompass soil, crops, and livestock are the beginnings of the modern study of farming systems. Thus the current emphasis on a holistic approach to agriculture does in fact have a very long history. The full span of a holistic overview must include the cultural aspects of agriculture, to be developed in the next chapter, but it should be clear already that systems thinking is one of the essential features of agricultural sustainability:

- Agriculture must be viewed and managed in a holistic manner.

The role of livestock was stressed in the previous chapter, but here we can see more clearly how they fit into the holistic overview. They have traditionally contributed energy for tillage, and they have provided manure for the maintenance of soil fertility. Thus they are key components both for planting crops and for keeping them growing well. Soil, crops, and livestock are essential and interdependent elements for agricultural sustainability. Although the energy requirement was noted in the last chapter, two more essentials now are clear:

- There must be a means of tilling the soil so that crops can be successfully grown.

To successfully grow crops, the seed must be placed in the soil so that it can germinate and the resulting crop plants are able to overcome the competition of weeds and the stresses of pests and bad weather. Tillage prepares the seedbed where germination can succeed, and it weakens and controls the competition that comes from weeds. Firming the seedbed allows good seed-soil contact. Even modern "no-tillage" methods that substitute mulches and herbicides for plowing still mix a small area of soil where the seed goes into the ground.

Not only is crop vigor dependent on the fertility of the soil, inadequate soil fertility can favor weeds (what the Bible often calls thorns) and even prevent the normal completion of crop growth and development. Grain will fail to form or will form only in small amounts. Because agriculture regularly exports nutrients in food for humans from the natural nutrient cycles in the fields where crops are grown, nutrient management is another essential of agriculture:

- There must be an effective means of maintaining soil fertility.

Without this, the cropping systems will simply run down and become unproductive.

## Conclusions

Systems are highlighted by thinking about the way soil, crops, and livestock interact and in fact are interdependent. The farming practices of soil tillage and nutrient management are examples of these interactions. The soil must be tilled so that crops can be sown and grown. Although agricultural systems developed in some places without livestock energy, the biblical account reveals a system in which the energy for tillage was provided by livestock. In addition, wherever they have been used, livestock have traditionally provided manure that fertilizes the soil and maintains its productivity. The concepts of integrated farming practices are not thoroughly developed in the Bible, but enough information is given so that we can identify three related essentials of agriculture:

- Agriculture must be viewed and managed in a holistic manner.

- There must be a means of tilling the soil so that crops can be successfully grown.

- There must be an effective means of maintaining soil fertility.

Even when tractors and manufactured fertilizers replace animals in agricultural systems, these essentials still apply.

# Chapter Nine

# The Culture of Agriculture

## Teaching and Learning

At one time I took a test to discover my "motivational abilities," and I learned that I was motivated to learn. That must be one reason I like to teach. Teaching is such an effective means of learning. In my efforts to be an innovative teacher, I have created situations where I have certainly learned new and unexpected lessons.

One of those situations occurred when I was developing a new unit for an introductory course in sustainable agriculture. The topic was cropping systems, and I was trying to illustrate comparative advantages and disadvantages. I had developed the lecture, and with the help of our teaching support specialist, we had established a field demonstration plot that included monocultures (crops by themselves) of corn, beans, and pumpkin growing beside a mixture of the same three crops together. The mixture is common in the traditional agriculture of the First Americans and is known among the Eastern tribes as the "Three Sisters."[1] A laboratory exercise was planned to harvest and measure the yield potential of the monocultures and mixture. To emphasize the point that we grow particular foods to meet nutritional requirements and maintain cultural traditions, I next looked for recipes that could be prepared for the class, recipes that incorporated all of the Three Sisters. That is where the process became very interesting.

It proved difficult for me to find the right kind of recipe. It was not that it did not exist, I just did not know enough as I began. Eventually, a colleague who was the director of the American Indian Program at Cornell put me in contact with the Oneida Tribe in Wisconsin. That led me to a collection of traditional recipes and finally a good starting point: "Indian corn soup," made with the same traditional white corn and dry beans we were growing in our plots. Unfortunately, the recipe was very traditional and called for "clean wood ash" to take the hulls off the corn and "salt pork" to add flavor. The contact in Wisconsin said I could simplify with reasonable results if I used canned hominy corn and bacon from the supermarket.

The next step was to see if I could actually follow the recipe with results suited to the classroom. So I experimented on my family. In a couple of tries and with a little added onion, the results were excellent. But it was not quite what I wanted. I wanted "Three Sisters Soup," and there was no pumpkin in the recipe. I decided to continue my learning process and just add the absent ingredient. I found some canned pure pumpkin at the supermarket and included a couple of tablespoons in my next test soup. Our middle son had liked the earlier variations and took the first mouthful of steaming Three Sisters soup. His face went blank, and then he spit the mouthful back into his bowl. Surprised, I took a bite from my bowl. The taste was satisfactory but the feel of the liquid in my mouth was like sand had been mixed in the soup. The pumpkin was full of what botanists call "sclerenchyma," small siliconized cells that are wonderful in pumpkin pie but like fine sand in a soup. That experiment had to be "recycled." There would be no more Three Sisters soup that year. As if that was not enough culinary learning, when I checked with my class I found that about a quarter of the students were vegetarians. So to stick with the plan, I had to create a second corn soup without bacon, but that is another story.

My effort to add food and culture to the lesson demonstrates the potential difficulties in trying to extend the boundaries of disciplinary learning. Those boundaries are defined, learned, and taught by professors like myself; but to study agriculture and agricultural sustainability, we must extend the disciplinary boundaries all the way to culture. After all, culture is the root of the very word that identifies the subject. The first part of the term *agriculture* comes from the Latin *ager*, which refers to the field, that is, the place where agriculture is practiced. The second part comes from the Latin *cultura*, which refers to care, cultivation, improvement, or refinement, that is, the way agriculture is practiced. One could argue that agriculture simply means the cultivation of the field, but that is too narrow. It misses too much of the context that must be included in a holistic appreciation of the subject. More than cultivation, the culture part of agriculture designates the span of the social elements accounting for the thoroughly human aspects of everything about food production. A list of prominent features would include economics, education, politics, and religion, as each relates to agriculture. A whole alphabet of topics could be constructed to cover the cultural aspects of food and farming, starting with anthropology and ending with xenoergonics,[2] which is pronounced with a "z." The interplay of the social sciences and the humanities with agriculture is profound and extensive. However, this chapter will not attempt to be comprehensive. Instead, it will continue the pattern of searching for the essentials of agriculture as they are revealed primarily in the ancient biblical writings. The social structure of the farm is one of the points of emphasis. In addition, we can discern various economic principles, for example, as they are related to ownership and labor, and to those religious aspects that are related to food, celebration, and moral behavior.

## Farmers in the Beginning

The Bible's first people, Adam and Eve (Gen. 2–3), were quintessential farmers according to theologian Theodore Hiebert.[3] God began his agricultural work by forming the first representative of the human race (*'adam* in Hebrew) from good farm ground (*'adamah* in Hebrew, Gen. 2:7). That first member of the human species was called Adam (Gen. 2:20), after the Hebrew name of the whole human race. According to the Bible account, the next agricultural work of God was planting a garden (Gen. 2:8). He then placed Adam in the garden "to work it and keep it" (Gen. 2:15, ESV). God the gardener and farmer made Adam a gardener and farmer. But the work of farming was not meant to be done alone. God recognized that Adam needed help (Gen. 2:18), so God brought the animals to Adam, and Adam gave them names (Gen. 2:19–20). To name (*shem* in Hebrew, to mark individuality) implied a studied familiarization. This acquisition of knowledge is the biblical foundation of science, of that aspect of human culture that studies the physical world.[4] Miller has argued that all human culture and not just science began at that point.[5] Out of his first study, Adam apparently identified from among the animals the species of livestock that became his literal companions and helpers in agriculture. The contributions of livestock and companion animals to human culture should not be underestimated. Beyond economic considerations, they have enriched our understanding and provided comfort in the nonmaterial aspects of human life.

But animal companions could not meet all of Adam's needs: [among them] "there was not found a helper fit for him" (Gen. 2:20, ESV). He needed another human companion, so, as the story goes, God formed the first woman from a part of the first man (from his side or rib) while he slept (under God's anesthesia, Gen. 2:21–22). When this familiar story is viewed from an agricultural perspective, we see that Adam and Eve were made to be a farm family. Their farming skills and property were passed on to their children (Gen. 3:16, 5:3–4). Though other cultures may have different models for the community of farmers, the biblical view is that the foundation of agriculture is the farm family. At the second beginning following the Flood (the story of Noah and the Ark), the farming mandate to humans was again affirmed:

> Noah began to be a man of the soil, and he planted a vineyard. (Gen. 9:20, ESV)

The Bible reveals a diversity of human occupations, but from the beginning and at the core, the essential human occupation is that of gardener-steward of God's creation. The Bible also stresses that parents are to teach their children, and that teaching includes their role as gardeners and farmers who steward God's creation:

*You shall teach* [all God's commands including those about farming] . . . diligently to your children, and shall talk of them when you sit in your house, and when you walk by the way, and when you lie down, and when you rise. (Deut. 6:7, ESV, emphasis added; see also Deut. 4:10, 11:19; Ps. 34:11, 78:5)

The biblical perspective shows God's concern for generations. The first covenant with Noah was for all future generations, as well as with all creatures (Gen. 9:12). God's blessings are said to reach a thousand generations (Exod. 20:6, 34:7; Deut. 5:10; Jer. 32:18), while the bad consequences of unfaithfulness only extend to the third and fourth generation (Exod. 20:5, 34:7; Num. 14:18; Deut. 5:9). God sometimes identified himself in generational terms as the God of Abraham, Isaac, and Jacob (Exod. 3:15–16, 4:5; 1 Chron. 29:18; Matt. 22:32; Acts 3:13 and elsewhere).

The Bible does not argue that the "family farm" is the only successful model for the farm enterprise. Probably the mixed cropping system that included livestock developed before the Bible was written down. As shown in the chapter on livestock (Chapter 7), dairy was an important part of that system. Agricultural historian David B. Grigg[6] reasoned that the family farm was a necessary aspect of dairy farming until recent times because dairy animals require more special attention than tenants are normally willing to provide. Thus "the land of milk and honey" may have been predisposed to the family-farm model. What the Bible does indicate in general is that there must be a means of education and generational transfer so that righteousness (right relationships with God, people, and land) is maintained.

## Land Tenure

Land tenure and property rights remain the subject of controversy and social experiment even today. In the West, the system of feudal lords in the Middle Ages evolved into the modern capitalistic system with private property and economic practices often contrary to biblical faith.[7] Experiments in socialism and communism have had varying amounts of success. However, the biblical pattern has seldom been applied, even among the ancient Hebrew people.

Two problems or abuses emerge with land and property. The first is that as societies become large and complex, property held in common is overused and undercared for. The "tragedy of the commons" is the standard ecological example in which the old common pastures were overgrazed to destructive levels because it always benefited an individual farmer to graze one more cow on the commons, even after the common pasture land was being overgrazed by neighbors who shared the resource. The other extreme is when an individual owner acts to destroy a resource that has value to neighbors but is not under their control. This problem may be amplified when the "individual" is a corporation where the goal is short-term profit, and property ownership is transferred not from generation to generation but from stockholder to stockholder. The argument that "this is my property and I can do with it whatever I please" is tempered when the property passes across generations to members of the same family. The Bible injunction against big farms, however that is defined, is clear:

> Destruction is certain for you who buy up property
> so others have no place to live. Your homes are built
> on great estates so you can be alone in the land. (Isa.
> 5:8, NLT)

A more literal translation of the same passage reads as follows:

> Woe to those who join house to house, who add field
> to field, until there is no more room, and you are
> made to dwell alone in the midst of the land. (ESV)

Perhaps there is divine intent behind the family-farming model just discussed because it inherently favors good land care across generations and stable numbers of people on the land, provided that farms are not progressively subdivided among multiple heirs.

Whatever interpretation is given to the biblical model of family farming, the Hebrew Bible is clear in laying down statutes for the inheritance of land. The first principle is that everything belongs to God (see Exod. 9:29; Ps. 24:1; Job 41:11). When the land promised to ancient Israel was finally divided among the tribes and families of the Hebrew people (Josh. 13:8–21:45), it was to be a permanent inheritance passing from generation to generation (as long as they obeyed the required laws):

> The inheritance of the people of Israel shall not be transferred from one tribe to another, for every one of the people of Israel shall hold on to the inheritance of the tribe of his fathers. (Num. 36:7, ESV; see also Ruth 2:20; Jer. 32:7)

Thus the land could be sold or rented, but not on a permanent basis:

> The land shall not be sold in perpetuity, for the land is mine. For you are strangers and sojourners with me. (Lev. 25:23, ESV)

Every fifty years[8] (the Year of Jubilee), the land was to be returned to the original inheritors (Lev. 25:10). The sale or rental price for the land was to be based on the number of harvests remaining until the Year of Jubilee, and the family selling land could buy it back at any time for 20 percent more than the assessed value (Lev. 25:15–16, 24–28, 27:19). The assessed value of land was to be based on its productive potential and the number of years remaining until the Year of Jubilee (Lev. 25:13–16, 27:16–18, 22–25). The questions asked in the original appraisal of the Promised Land included these:

> How is the soil? Is it fertile or poor? Are there many trees? . . . [B]ring back samples of the crops you see. (Num. 13:20, NLT)

Even land dedicated to God and used by the priests was to be offered for sale to the original owners in the Year of Jubilee (Lev. 27:17–21). There appears to be a loophole in the law that allowed priests to acquire lands from others who could not afford to buy it back in the Year of Jubilee. However, Nehemiah 5 (verses 5, 10–11) corrects this abused privilege. Stealing land was a very serious crime:

> . . . [N]ever steal someone's land by moving the bound-
> ary markers your ancestors set up to mark their prop-
> erty. (Deut. 19:14b, NLT; see also Exod. 20:15, 17; Deut.
> 27:17; Mic. 2:2; Luke 12:15)

The biblical pattern for property rights and responsibilities solves some of the problems of both private and public ownership mentioned earlier. It is always clear just which property is an individual's or a family's responsibility. In this sense, there is no commons. The property also has clear and fair value so that someone who cannot farm it can still gain a living by selling or renting it. Thus there is a kind of social security based on land that was to apply to everyone in common. Poverty and misfortune were not supposed to drive a family permanently from its land heritage. There was a way for future generations in a family line to get back into farming.

## Labor and Financial Matters

Though the law of the Hebrew Bible provided for restoration every Jubilee, misfortune and poverty did occur and also were subject to biblical guidelines. Tenant farmers were to pay the land-lord one fifth of their crop as rent (Gen. 41:34, 47:23–24; in these cases, the landlord was the Pharaoh of Egypt). Thus the Bible sanctions a kind of share cropping. Hired workers were to be paid promptly (Lev. 19:13; Deut. 24:14–15; Mal. 3:5; James 5:4). Workers as well as livestock working in a crop were to be allowed to eat of the fruit of their labors (Prov. 27:18; 1 Cor. 9:7, 10). The biblical picture is of a community of participants in the farming system, owners, workers, and their livestock, all of whom were to receive fair and benevolent consideration. Fair treatment also extended to consumers. A farmer with a crop to sell was not to hold it for a higher price while his neighbors suffered for want of food (Prov. 11:26; Neh. 5:1–5, 10–11), and when grain and other goods were sold, the sale was to be honest:

> [35]"Do not use dishonest standards when measuring
> length, weight, or volume. [36]Your scales and weights
> must be accurate. Your containers for measuring dry

goods or liquids must be accurate. (Lev. 19:35–36a, NLT; see also Deut. 25:13–16; Prov. 20:10; Amos 8:5–6)

Of course debts were to be paid back (Lev. 25:27; Ps. 37:21), but when one in financial need had to borrow money or become indentured labor, one was to receive good terms and could not be held in debt for longer than seven years (Deut. 15:1–2, 31:10). A borrower's means of livelihood was never to be held as collateral (Exod. 22:26–27; Deut. 24:6, 13, 17; Ezek. 18:7). The goal was that there be no poor among them (Deut. 15:4). However, that is a goal never fully realized (Deut. 15:11; Matt. 26:11), so interest was not to be charged to the poor.

> If you lend money to any of my people with you who is poor, you shall not be like a moneylender to him, and you shall not exact interest from him. (Exod. 22:25, ESV; see also Lev. 25:35–37; Deut. 23:19–20; Neh. 5:11 and elsewhere)

Foreigners could be charged interest. The Bible provides detail for a just economic system that goes beyond what can be covered here, but in summary it provides an ancient view of what constituted economic morality.

The biblical goal of economic morality was not only the minimization of poverty but also that people would prosper through their right relationship with God:

> Therefore keep the words of this covenant and do them, *that you may prosper* in all that you do. (Deut. 29:9, ESV, emphasis added)

> [1]Blessed is the man who walks not in the counsel of the wicked, . . . [2]but his delight is in the law of the Lord, and on his law he meditates day and night. [3]He is like a tree planted by streams of water that yields its fruit in its season, and its leaf does not wither. *In all that he does, he prospers.* (Ps. 1:1–3, ESV, emphasis added)

D. L. Miller sees a "development mandate" in Genesis 2:15 through which natural resources are to be developed for the purpose of bringing glory to God and prosperity to humans by their care and by expansion of the garden they have been given to steward.[9]

These few passages show that the biblical plan is not just for glory after death but also for a sustainable future for people and their descendants in this life. The promises are conditional, and hence much religious or theological effort has gone into detailed analyses found in many other writings. However, enough has been said to show that the divine intention reported in the scriptures is for farming, as just one of the many occupations, to provide blessings and prosperity to farmers as they follow God's plan for relationships with him, with one another, and with the land.

## Food and Culture

Perhaps one of the prime distinguishing characteristics of human cultures is the food that is eaten, especially for celebrations. That aspect of ancient Hebrew culture, as far as I can discern, does not have universal dietary applications (though some will disagree), but it certainly illustrates the general point for a specific case. For example, "milk and honey" are mentioned twenty times together in the Hebrew Bible. Roasted lamb was prominent in the Passover feast (Exod. 12:21; Ezra 6:20; Mark 14:12), along with unleavened bread (matzo) and bitter herbs.

Because human cultures have emerged in different environments, it is not surprising that there is great diversity among cultural foods. This extends not only to the ingredients but also to the way they are prepared. For example, cheese is mentioned in the Bible (twice, ESV) but not ice cream. Fish are mentioned fifty-eight times (ESV) but not canned tuna. The general food processes of preparation and distribution also have unique biblical examples: Grain was parched (mentioned eleven times in the ESV), grape juice was made into wine (mentioned 212 times), and because of lack of safe storage, meat was to be eaten soon after it was prepared (Exod. 16:8; 1 Kings 17:6). Cultural feasts often were associated with celebrations of thanksgiving and praise.

> When you have eaten your fill, praise the LORD your God for the good land he has given you. (Deut. 8:10, NLT)

Leviticus describes how food was to be used in thanksgiving offerings (Lev. 7:12–13, 15, 22:29). An important element of the Christian celebration of the Lord's Supper was remembering that Jesus thanked God for bread and wine (Matt. 26:26–27; Mark 14:22–23; Luke 22:17, 19; 1 Cor. 11:24).

## Ethics and Morality

Religion is one of the great cultural constructs of human society, and one of its functions is to define and promote ethical or moral behavior. Our personal, functional religion is the perspective that informs our understanding of what we ought to do, and knowing what we ought to do is at the heart of designing new or improved methods of agricultural sustainability. The goal of these chapters has been to get our value systems and worldviews into the discussion of agricultural sustainability. Certainly the Judeo-Christian writings say much about moral behavior. Libraries of references are available to those who wish to delve deeply into the subject. However, the biblical standards for moral living can be summarized in just a few passages:

> He [God] has told you, O man, what is good; and what does the LORD require of you but to do justice, and to love kindness, and to walk humbly with your God? (Micah 6:8, ESV)

> [29]Jesus answered, "The most important [commandment] is, 'Hear, O Israel: The Lord our God, the Lord is one. [30]And you shall love the Lord your God with all your heart and with all your soul and with all your mind and with all your strength.' [31]The second is this: 'You shall love your neighbor as yourself.' There is no other commandment greater than these." (Mark 12:29–31, ESV; see also Deut. 6:4–5; Lev. 19:18; Matt. 22:37; Luke 10:27; Gal. 5:14; Jas. 2:8)

> So whatever you wish that others would do to you, do also to them, for this is the Law and the Prophets. (Matt. 7:12, ESV; see also Luke 6:31)

The simple summary is that we are to love God completely, and we are to love our neighbors as ourselves. The definition of love goes beyond our purposes here (but see 1 Cor. 13). There is a biblical moral foundation for loving our neighbors, who benefit or suffer from our ways of obtaining food.

## The Essentials

Looking first at the lifestyle of farming, it is clear that ancient wisdom stresses that there be a means of teaching new farmers (generational transfer), and that life on the farm involves a kind of prosperity and a means of living so that people will continue to want to farm. The family farm stands out as particularly important, but perhaps other means of training and transferring the resources of farming across generations could achieve the basic goal of maintaining a population of farmers who will know how to farm in their particular environment and who will want to farm because it represents an attractive lifestyle. This is the essential point:

- Farming must offer an attractive lifestyle and a means of learning how to farm so that future generations will become farmers.

Another principle that emerges from the ancient writings as they relate to culture is the concern for justice and equity. There is a moral or an ethical element that is practically related to the well-being of one's neighbors, including neighbors from future generations. The economic and social system should take care of people so that the disadvantaged do not starve. The Bible describes and endorses economic activity and development, which can be regarded as the stewardship of resources suited for development. But it limits economic development by concerns for those who are at a disadvantage as the development takes place. The financial goals of the wealthy should be tempered by helping provide the unfortunate with the basics of life, including food provided by agriculture and the opportunity to make an economic and a social recovery as provided in the Year of Jubilee.

- There must be an economic system that rewards stewardship of resources and provides a way for the disadvantaged to be fed and to recover.

This point should be elaborated with a corollary related to justice for all participants in the food system, including farm laborers, the poor, and livestock. The biblical view is that any human endeavor, including agriculture, must be moral and ethical. As a minimum, that has the following meaning:

- There is a religious and an ethical component of agriculture that calls for all participants to have food so that they can celebrate life.

Finally, taking this chapter and the previous one together, we now have at least a sketch of the dimensions of systems thinking about agriculture. The agricultural system has both biological and physical components and social or cultural components. Our earlier conclusion stands: Agriculture must be viewed and managed in a holistic manner. However, we now perceive that holism is quite far-reaching. Religious and ethical issues are as much a part of the tapestry of agricultural sustainability as are economics and farm labor issues or the selection of crop varieties and tillage machinery. The participants in the system also stand out more clearly than before. All people who eat are a part of the system, but those struggling to obtain food warrant special attention. The holistic perspective also makes farm livestock and the wild species impacted by farming part of the system. Agriculture can be seen as an essential aspect of human culture that generates food, not only for sustenance but also for celebration of life and representation of meaning.

## Conclusions

The culture of agriculture is more than the methods of farming. It is the whole fabric of human activity that ultimately depends on the food system for life and meaning. This very brief overview of a very large topic has brought out three more essentials of agriculture:

- Farming must offer an attractive lifestyle and a means of learning how to farm so that future generations will become farmers.

- There must be an economic system that rewards stewardship of resources and provides a way for the disadvantaged to be fed and to recover.

- There is a religious and an ethical component of agriculture that calls for all participants to have food so that they can celebrate life.

In addition, the previous essential (agriculture must be viewed and managed in a holistic manner) has been clarified in terms of how much it actually covers. To think holistically, we must include everything that is important. Religion or worldview is important because it is the foundation of motivation for change. To become more sustainable, we are going to have to change, and that will involve getting in touch with what really motivates people.

# Chapter 10

## Abuse, Poverty, and Women

Abuse is wrong use. When it comes to people, abuse is using persons. The rhetoric of agricultural sustainability is about abuse. The environment is abused, livestock are abused, farmers are abused, consumers are abused, and so on. Whether we agree or disagree regarding the extent of abuse, the idea is that as agriculture becomes more sustainable, there will be less abuse, in qualitative and quantitative terms. Thus reducing or avoiding abuse is a part of the definition of agricultural sustainability.[1] Agricultural sustainability will be defined in detail in the last chapter (Chapter 12), but in the present context, we can say that it involves meeting the present and future needs of all people for food and health while caring for their natural and social environment. In the preceding chapters, we have used the Bible to compare ancient and modern perspectives in order to discern, as best we can, what has to be done in agriculture and how well we have been doing it. A list of the essentials of agriculture has been completed (Appendix 3), but there are still some dimensions of abuse that warrant closer examination. Early in the twenty-first century, abuse of the poor and abuse of women are interrelated problems that affect agriculture. Recent perceptions and ancient biblical perspectives are again combined in this chapter because they enrich our understanding, giving more insight than is possible with either alone.

## Poverty, Women, and Agriculture

The current association of these factors has been attracting special notice, especially in Africa, where up to 80 percent of the agricultural labor force in some countries is made up of women.[2] Over 1.3 billion people in the world live in absolute poverty, with each new day bringing the challenge of obtaining the food needed for survival. Often, landowners (who are comparatively rich) legally force the poor from the good agricultural land. In order to grow food, these poor people must use marginal land in fragile environments that cannot support agricultural production for very long.[3] The result is often severe soil erosion and the rapid deterioration of land and water. Often the land is also deforested, not only to make room for crops but also to obtain firewood. Where necessary, livestock manure is used for fuel instead of fertilizer for the land. Poverty is the driving force behind such abuse of natural resources. The sad fact is that poverty is becoming more and more associated with women. More than two out of three of the world's 1.3 billion poor are women.[4] More men than women have gone to the cities to look for work, and simply for cultural reasons, some men expect women to be the providers of water, fuel, and food for the household. The worldwide feminization of poverty and the association of poverty with environmental destruction are big concerns for all who seriously consider the future of the human race and the environment.

The problems are of course exceedingly complex.[5] Population issues are intertwined and controversial. Associated racism and gender bias have cultural dimensions that complicate solutions. One study wisely argued that all people need to be empowered to be good stewards. Efforts that focus on women alone will not be adequate.[6] Literacy and education are generally accepted as important contributors to solving the related problems, both as they affect human reproduction and environmental care. One of my mentors once told me that though there is no data, he thought that literacy for a farmer would increase production by about 10 percent. How much more literacy adds to the quality of life! In many places women have been selectively denied education, and the literacy rate among women is about half that of men in developing countries such as Ghana in Africa. The situation is improving in Ghana, where almost half of the primary school students are now women. Women also have been organizing to gain recognition and a voice for their role as farmers.[7] However, the question remains whether enough is being done to avert further disaster and escalating human suffering.

At the heart, these problems are moral or ethical. As pointed out in the previous chapter, the Bible provides a strong moral basis for caring for the poor, both in meeting immediate needs and in offering hope and guidance for improving their situation. The biblical material is exceedingly rich and the subject of many other books. Two good examples are by Sider[8] and by Blomberg.[9] The same can be said for the study of women in the Bible. I have found contributions by Witherington[10] and by Sumner[11] particularly helpful. It is doubtful that I can do better than send the reader in search of these and other relevant contributions. However, two aspects of the subject I have not seen addressed elsewhere prompted me to write this chapter. In the context of agriculture, and specifically the sustainability of agriculture, what are the dimensions of these problems that must be considered as we look for holistic and lasting solutions? As we envision, design, and implement processes for improved sustainability, we at least need checklists for evaluating critical impacts on the poor and on women. Before developing example checklists from the Bible, a brief personal reflection will help demonstrate the complexity of the issues involved here.

## A Personal View of Poverty

Our personal and inevitably biased perceptions of poverty come from our experiences and social context. For me, the first awareness of poverty was interpreted by my parents. Both matured in the Great Depression of the 1930s, and both were children of tenant farmers. My father was a stepchild whose stepfather lost his property through a series of bad economic decisions. My father's frugal mother nurtured their resources so that Dad inherited a few cows when his mother died in 1941. My mother had no inheritance at all. By today's standards in the United States, they were poor, but I do not think they seriously thought of themselves as poor.

I remember asking my parents once if we were poor. I was probably puzzled by the introduction of the concept at school, perhaps when I was in the fourth grade. I recall that their response was to laugh and reassure me that our family was not poor. Then they got into a good-natured discussion about which of them was the poorest during the Great Depression. Mother said her family was so poor that they could not afford to eat ordinary corn when she was growing up. They had to eat hominy, which is a corn grain by-product with the nutritious germ of the seed removed. Dad said he thought his family was even poorer. They made butter to sell in town, so they were not allowed to eat butter on their own bread at home. Instead, they spread their bread with lard, the boiled-down fat of a pig. Then they agreed that the poor family they knew about had twelve children and could not afford to buy oatmeal for breakfast during the winter. Instead, they got a less-expensive wagonload of cracked wheat. For them, that was real poverty. Both of my parents also had served as nurses for sick and dying family members because they could not afford adequate medical care. Still, they did not seriously count themselves among the poor. When I was young, they were realizing a dream of building a ranch, and in the economic boom following World War II, our family's economic status improved every year.

Their reassurance did not stop my puzzlement over poverty. The following summer my toes wore through the ends of my shoes. I asked my parents if I might get a new pair of shoes. Their answer was to wait until it was time to buy clothes for school. It was alright to have holes in your shoes. Besides, my mother said she regularly went barefoot when she was growing up because shoes were too expensive then to wear all of the time. Behind their response was the fact that my parents were buying a ranch, and there were land payments to be met.

Our economic situation improved greatly, so much so that with the help of public-supported scholarships I was able to go to college. It was as a college student that I became a Christian and joined a church. And it was in the Christian context that I was exposed to the civil rights movement of the 1960s, with its understanding of oppression, social injustice, and lack of education as the root causes of poverty. I then married into a family with missionary commitments to relieve poverty in other places through work in international agriculture and economic development. They also were religious pacifists, and they saw massive military expenses as a misallocation of economic resources that could have helped needy people. Because of them, I traveled in Central America in the 1970s and encountered First Americans so poor that they never had shoes. However, these people probably did not count themselves as poor either. They certainly were rich in relationships and culture, and they were very generous with what they did have.

The point of this personalized view of poverty is to illustrate the complexity of the subject. Our own experiences cannot reveal everything, so we come with biases and blind spots. Recognizing poverty also is so relative. From my experiences in Central America, I saw that just about all Americans and Canadians are rich in comparison. At the same time, even the very poor did not always identify themselves as such. So the perceptions of poverty are comparative, relative, and necessarily biased through limited experience. The compiled ancient wisdom of the scriptures can help us reach a more comprehensive understanding. Such a broad perspective should improve the evaluation of our efforts at sustainability and the related problems of poverty.

## Proverbs on Poverty

Mark Graham makes the point that there is a Christian moral mandate that our agricultural systems do not "cause or exacerbate poverty and the vulnerability, marginalization, and powerlessness associated with it."[12] This is a fact well supported by Scripture. There is so much in the Bible about care for the poor that the direct and indirect references seem to be almost uncountable, in part because of our personal differences in identifying what is relevant. However, there can be no doubt that the Bible teaches that God cares for the poor, and that he wants us to do likewise:

> LORD, who can compare with you? Who else rescues the weak and helpless from the strong? Who else protects the poor and needy from those who want to rob them? (Psalm 35:10b, NLT)

> [D]o not oppress the widow, the fatherless, the so-
> journer, or the poor, and let none of you devise evil
> against another in your heart. (Zech. 7:10, ESV)

> As the Scriptures say, "Godly people give generously to
> the poor. Their good deeds will never be forgotten." (2
> Cor. 9:9, quoting Psalm 112:9, NLT)

Although the whole Bible contains a great deal of information about the plight and the just treatment of the poor, one relatively small book of collected wisdom provides a fairly comprehensive listing of the causes of poverty. The book of Proverbs, which is a collection of instructions on how to live wisely, identifies twenty causes of poverty. The list that follows, which is based on those proverbs, thus gives twenty checkpoints for judging the possible impacts of plans for sustainability on the poor.

At the outset, we should note the unique perspective of the book of Proverbs, which is focused on personal instruction and emphasizes responsibility for one's own condition. Though societal responsibilities are addressed, they are mixed with the personal responsibilities not emphasized in a modern sociological analysis. In addition, some of the ideas are incomplete in themselves and need to be placed in the context of the full list in order to be correctly understood. Given that perspective, the twenty proverbial causes of poverty are enumerated here, in the order of occurrence. Again, the objective is to compile a checklist that can help us evaluate plans for sustainability in terms of the possible effects on the poor.

*1. Immoral sexual behavior.* This is the first cause of poverty noted in the book of Proverbs. Immorality often leads to failure of marriages, broken homes, and solo parents and providers. Sexually transmitted diseases are increased by marriage infidelity. Although the spiritual aspects of immorality are emphasized by religion, there are clear economic consequences as well. In this context, the male is addressed:

> [8]Run from her [the immoral woman]! Don't go near the
> door of her house! [9]If you do, you will lose your honor
> and hand over to merciless people everything you have
> achieved in life. [10]Strangers will obtain your wealth, and

someone else will enjoy the fruit of your labor. (Prov. 5:8–
9, NLT; see also Prov. 2:16, 6:24, 23:27–28)

The economic hardships resulting from broken marriages and no
marriages at all are well known in the United States. Single parents need special social and economic support along with understanding and compassion, but as we seek solutions to these
problems, we need to address the moral as well as the economic
issues. Stated in the most general terms, sustainable agricultural
practices should strengthen families and encourage the faithful-
ness of spouses.

    *2. Laziness.* This is a second cause of poverty identified in the
book of Proverbs.

Lazy people are soon poor; hard workers get rich. (Prov.
10:4, NLT; see also Prov. 12:24, 13:4. The ESV translation of Prov. 10:4 reads as follows: A slack hand causes
poverty, but the hand of the diligent makes rich.)

This first appears to be an oversimplified, shallow treatment of
poverty, but a deeper meaning is implied in this verse. An analysis
of the original Hebrew wording shows that this is not a simple-
minded overemphasis on the work ethic. To start with, most translations indicate that the kind of laziness addressed here is related
to unwillingness to work with one's hands. The work ethic is important, as emphasized by point 7 that follows, but that is not the
full meaning here. The original Hebrew for this verse uses the
term *rem·ee·yaw'*, which means deceitfulness as well as laxness. Thus
most English translations communicate only part of the original
meaning. The original meaning denotes two ideas. One is a de-
ceitful laziness that does not deliver what is paid for. That is the
implication of the earlier NLT translation. The other is the laziness of accepting simplistic and false interpretations of a problem
situation, a kind of self-deception that comes from not doing the
work of getting to basic causes that must be addressed in order to
reach lasting solutions for a problem. Ascribing all poverty to
laziness is an example of an overly simplistic analysis that will not
solve the underlying problems. There are at least twenty causes of
poverty, not just one.

The passage also seems to say that hard work will always make one rich. That too is a simplistic and an inaccurate interpretation corrected later. The term for hard work used here is *khaw·roots'*, which means diligent and sharp. This implies consistent and persistent work that will penetrate to basic causes in balanced contrast to the laziness denoted by *rem·ee·yaw'*. This passage also is about the ideal that hard workers should be rewarded. Later we learn that because of injustice (point 6) and oppression (point 8), hard work may not lead to wealth. Some may argue that the riches of Prov. 10:4 are "riches of the soul" that are more important than material wealth. That also may be true, but I think most would agree that when one works hard to make a living, there should be corresponding material benefit; but sometimes there is not. One must conclude that something is wrong with a social system that does not deliver what is earned. A sustainable approach should reward hard work. Applied to sustainability, this verse indicates the need for a thorough and an accurate analysis of the problem as well as diligent work habits matched by fair rewards.

*3. Stinginess.* An aversion for sharing is a third cause of poverty noted in the book of Proverbs.

> It is possible to give freely and become more wealthy,
> but those who are stingy will lose everything. (Prov. 11:24, NLT)

This idea is amplified in the teachings of Jesus (Luke 6:38) and of Paul (2 Cor. 9:6; Gal. 6:7), where the implication is that poverty comes to the person who does not give. It follows that the best practices of sustainability will foster generosity, both in the wealth of ideas and of material things to share. A sustainable approach should encourage sharing and provide an abundance to give away.

*4. Wasting time and land.* There is a biblical association of waste and foolishness that results in poverty and hunger.

> Hard work means prosperity; only fools idle away their time. (Prov. 12:11, NLT; see also 28:19; in the ESV translation of Prov. 12:11, the verse reads as follows: Whoever works his land will have plenty of bread, but he who follows worthless pursuits lacks sense.)

The ESV translation clearly links food poverty to the failure to use the land resources we have to produce food. The hard work mentioned here is *abad*, which is the service of stewardship described in Chapter 2, where it also is applied to the care of the land. To idle away time (or to follow worthless pursuits) is the translation of a Hebrew term, *rake*, for being empty like a poured-out jug. Overcoming food poverty is thus linked in Scripture to the combination of accepting a stewardship responsibility while rejecting empty expenditures of time and energy. This idea fills out the concepts of hard work and laziness, discussed in point 2. What is addressed here is the importance of the care or service of stewardship as a component of a sustainable system.

5. *Failure to change.* Not responding to instruction or criticism can perpetuate the downward spiral of poverty. An agricultural example would be a farmer who says, "I grow tobacco, so I will not try vegetables," or, "I raise cattle, so I will not raise sheep" when the new enterprises might improve the farmer's situation.

> If you ignore criticism, you will end in poverty and disgrace; if you accept criticism, you will be honored. (Prov. 13:18, NLT; the NASB says that poverty comes to the person who "neglects discipline"; see also 13:8.)

Unwillingness to change is linked to pride in point 20 that follows. There it is the pride of being unwilling to give help. Here it is the pride of being unwilling to seek and accept help. Applied to sustainability, we want approaches that provide and encourage ongoing critical evaluations with practical options for change. This fills out the concept of disciplined, accurate analysis, emphasized in point 2, but it adds the implication of critical participation by others, of getting help from others.

6. *Injustice.* The absence of social or economic justice prevents the farmer (or anyone else) from benefiting from his or her own hard work.

> A poor person's farm may produce much food, but injustice sweeps it all away. (Prov. 13:23, NLT)

One form of injustice is a lack of fairness that keeps the poor from prospering. Sustainability in agriculture must foster social justice that provides food, opportunity, and rewards for personal efforts to improve life's circumstances.

7. *Mere talk.* This is idle talk. Some work involves talking, but talking without working is a kind of laziness.

> Work brings profit, but mere talk leads to poverty! (Prov. 14:23, NLT)

This verse means that we need to do more than just talk about agricultural sustainability. Plans must be put into practice. There is work to be done and practices to be followed. The Hebrew term for work used here is *eh'·tseb*, which means labor and toil. This contrasts with *khaw·roots'* (point 2), which implies a sharp diligence, and *abad* (point 4), which denotes care and stewardship. All three of these Hebrew concepts of work are related to overcoming poverty. The implication is that getting out of poverty and achieving effective sustainability will take diligent and persistent effort and care, what is often called "a good work ethic."

8. *Oppression.* Synonyms for oppression include domination, coercion, cruelty, and tyranny. Whichever term is used, oppression of the poor by the powerful is degrading and tends to keep them poor.

> Those who oppress the poor insult their Maker, but those who help the poor honor him. (Prov. 14:31, NLT; see also 17:5)

Relieving oppression, especially by the powerful and rich on the weak and poor, should also be a part of agricultural development plans that will be truly sustainable. This proverb illustrates the ancient belief that God cares about the poor, and for us to do otherwise is an insult to God.

9. *Procrastination.* Failure in timeliness can have serious economic consequences in many situations, but in agriculture it often results in poor yields and little food.

> If you are too lazy to plow in the right season, you will have no food at the harvest. (Prov. 20:4, NLT)

For the experienced farmer, this is just common sense; but again, as we design new practices for sustainability, they must not interfere with the commonsense essentials of agriculture, or they will not succeed. The concept of laziness also appears in the previous point 2, but here it is slightly different, coming from the Hebrew term *aw·tsale'*. This kind of laziness means idleness or the absence of work, which is the usual meaning of laziness in English. One dimension of the kind of laziness meant here is the laziness of not looking after the details. This is clarified and amplified in point 10.

*10. Lack of attention.* Inattentiveness is another problem related to poverty. Figuratively, it is not keeping our eyes open. Loving sleep is used as a metaphor about the problem.

> If you love sleep, you will end in poverty. Keep your eyes open, and there will be plenty to eat! (Prov. 20:13, NLT; see also 19:15, 23:21b)

> [10]A little extra sleep, a little more slumber, a little folding of the hands to rest—[11]and poverty will pounce on you like a bandit; scarcity will attack you like an armed robber. (Prov. 6:10–11, NLT; see also Prov. 24:33–34)

Another proverb (Prov. 3:24) addresses the benefits and blessings of sleep, so clearly that is not the point here. Instead, this proverb reminds us to pay attention to what is going on when we should be awake. The book *Fast Food Nation*[13] describes far-reaching changes that occurred in the American food system over fifty years, from about 1950 to 2000. Many of those changes had undesirable effects on poverty as well as on human nutrition. Had more people been awake to the consequences of the gradual changes that were occurring, perhaps some of the problems could have been avoided. Part of an effective plan for sustainability will be a wakeful and continuing appraisal of the consequences of the changes being made.

*11. Hastiness.* This proverb reminds us that lack of planning followed by hasty shortcuts can have undesirable consequences. The prudent and successful farmer knows the seasons and plans ahead.

> Good planning and hard work lead to prosperity, but
> hasty shortcuts lead to poverty. (Prov. 21:5, NLT)

Good planning is here linked to the hard work of diligence
(*khaw-roots'*), also mentioned in the previous point 2. The application of the underlying principle in agricultural sustainability concerns both the need for thorough planning and the manner in
which the actual work of farming is done. One can argue that not
everything must be done well, but the essentials require thor-
oughness and timeliness, which add other dimensions to the concept of hard work, introduced in point 2.

*12. Lack of compassion.* Suffering can be ignored or compas-
sionately addressed. Ignoring poverty not only fails to solve problems, it can actually make them worse.

> Those who shut their ears to the cries of the poor will
> be ignored in their own time of need. (Prov. 21:13, NLT)

One of the points of this short verse is that pain and suffering
have a way of spreading in a broken system. Agriculture must
successfully provide for the basic needs of food and health, or it
will not be sustainable. Lack of food and health is a symptom of
the most severe kind of poverty. The Judeo-Christian view is that
existence of these problems should be met with practical compassion that attempts to meet the immediate needs and to correct
the structural problems in the food and health systems that lead
to those problems in the first place.

*13. Luxurious living.* Extravagant spending of resources on
luxuries is to be avoided.

> Those who love pleasure become poor; wine and luxury
> are not the way to riches. (Prov. 21:17, NLT)

The idea here is that luxurious living contributes to the problems
of poverty. Obviously those who indulge in luxury when they cannot afford it are making themselves even worse off (see point 14),
but those who can afford luxury engender a kind of spiritual
poverty with self-centered indulgence. This may be the opposite
of stinginess (point 3), but it produces some of the same results.

Luxurious waste not only drains away resources, it keeps some resources from reaching the poor. A modern proverb admonishes that we should live simply so that others can simply live. Simple living is a virtue of sustainability because it works against wasteful extravagance. We should note, however, that sustainability fosters diversity and local adaptation so that sustainable systems often are involved rather than simple. Living simply does not mean that we live in a simple system.

*14. Unwise borrowing.* Borrowing in general is part of the poverty syndrome.

> Just as the rich rule the poor, so the borrower is servant to the lender. (Prov. 22:7, NLT)

Debt management is a crucial aspect of getting in and staying in agriculture and must be facilitated by a sustainable system. As an example, back in the 1980s in the United States, there was much agricultural borrowing based on inflated land prices. When land values fell, many farmers became too poor according to the financial system to continue to farm, and they had to leave their farms for other work. A sustainable system will minimize the stress and abuses associated with debt.

*15. Exploiting people.* Exploitation can be by force or manipulation, but either way it has bad economic consequences.

> A person who gets ahead by oppressing the poor or by showering gifts on the rich will end in poverty. (Prov. 22:16, NLT)

The histories of slavery in the United States and of the more recent development and failure of many communist economic systems demonstrate this principle. A sustainable system will minimize exploitation either by force or manipulation. It will not reward the rich at the expense of the poor.

*16. Addictions and uncontrolled appetites.* Addictive behaviors can cause poverty and complicate recovery.

> [20]Do not carouse with drunkards and gluttons, [21]for they are on their way to poverty. (Prov. 23:20–21a, NLT)

Uncontrolled addictions lead to economic and health disasters wherever they occur. The growing list of modern addictions includes eating disorders, and those disorders can be linked to our modern food system. Addictions often are coping mechanisms for dealing with personal stress. Sustainability will alleviate instead of exacerbate human stress in all parts of the food system, from the farmer and land steward to the ultimate consumers who provide food for themselves and their families.

*17. "Poor-on-poor" oppression.* The book of Proverbs specifically identifies this kind of oppression, proving that it has been going on since ancient times.

> A poor person who oppresses the poor is like a pounding rain that destroys the crops. (Prov. 28:3, NLT)

The generalized meaning of this agricultural analogy is that those who should help because they face a common problem can instead become a source of harm. Oppressive systems often divide and conquer, pitting one disfavored group against another. Sustainability, in contrast, will foster cooperation and generosity between participating groups, and though competition may be encouraged, it will be limited so as not to become socially and culturally destructive. Moral constraint will characterize a sustainable economy.[14] Point 19 further develops this concept.

*18. Corrupt government.* A government sets policies and practices, and when the government is wicked or corrupt, life for the poor becomes even harder.

> A wicked ruler is as dangerous to the poor as a lion or bear attacking them. (Prov. 28:15, NLT)

The implication is that just government will benefit the poor. The government is able to do things to alleviate poverty and promote sustainability that individuals and private groups cannot do. General regulation and enforcement are probably best done by governments. On the other hand, governments also can get involved in other things better handled by individuals or nongovernmental groups. Fostering personal relationships that promote character change is an example. Proper governmental involvement is a challenging but an important aspect of developing sustainability in agriculture. Ikerd discusses at length the role of government in sustainability.[15]

*19. Greediness.* This is the underlying abuse behind several other factors in this list. Stinginess, injustice, oppression, hastiness, lack of compassion, and exploitation all have roots in greed.

> A greedy person tries to get rich quick, but it only leads to poverty. (Prov. 28:22, NLT)

The greedy person may or may not be the one who becomes poor, but the consequences of greed can be seen in many of the sicknesses in society. One of the troubling points about economic competition and the free-market economy is the underlying acceptance of greed as the primary human motivation. The economic approach by itself does not account for all that is desirable and even essential in human relationships, so a sustainable approach will recognize and provide for other factors such as human generosity and concern for the future.[16, 17]

*20. Pride and disdain.* What is meant by pride and disdain is people's self-centered attitude and not caring about others. Arrogance and conceit are synonymous words.

> [13]They [the selfish] are proud beyond description and disdainful. [14]They devour the poor with teeth as sharp as swords or knives. They destroy the needy from the face of the earth. (Prov. 30:13–14, NLT; the context is in reference to those who are never satisfied, who always want more; see also 1 Tim. 6:9–10)

One aspect of the modern drive for continuous economic growth is an underlying state of never being satisfied. Ikerd calls it a destructive social cancer.[18] The hallmark of pride and disdain is an unwillingness to change such patterns. Though twentieth on this list, which is based on the order of mention in the book of Proverbs, some people who are experienced in social and cultural efforts to alleviate the problem of poverty would rank pride and unwillingness to change as our foremost human problems. Pride is repeatedly mentioned as a human failing in this book (Prov. 16:18, 21:24, 29:23), and elsewhere in the Bible, pride is linked to failure to care for the poor and needy (Ezek. 16:49). At the heart of concern for sustainability is the recognition that we need to change our selfish ways so that our environment receives better care and our companions are fairly treated. This proverb makes it clear that selfish pride is destructive.

This list of proverbs on poverty is notable for two reasons. The first is what it contains. A modern analysis of poverty would include many of the same factors as this ancient list that is perhaps 3000 years old. The second is what it does not contain. Lack of knowledge and lack of education are not listed as causes of poverty. Is that an oversight, or is it profound insight? Certainly instruction, knowledge, and wisdom are highly valued in the Bible. Here is just one example:

> [10]Take my *instruction* instead of silver, and *knowledge* rather than choice gold, [11]for *wisdom* is better than jewels, and all that you may desire cannot compare with her. (Prov. 8:10–11, ESV, emphasis added)

Most of the items on the ancient list are related to abuses and attitudes that lead downward. Education is different, in that it is part of the solution of the poverty problem and leads upward. Knowledge and instruction will be ineffective unless the underlying abuses and attitudes also are corrected.

For readers wishing to see a more comprehensive but still relatively short presentation on biblical principles about the treatment of the poor, Michael Oye's chapter in *Biblical Holism and Agriculture* is useful.[19] He summarizes the laws of the Hebrew Bible that help correct the problems of poverty: daily payment of wages (Lev. 19:13), giving of a tenth of agricultural production as charity (Deut. 14:28–29, 26:12), interest-free loans (Lev. 25:35–37), right of the poor to glean produce that has been left behind (Lev. 19:9–10), restoration of property sold or mortgaged (Lev. 25:25–30; Deut. 15:1–4, 11), and equal participation in feasts (Deut. 16:16–17). Every point relates to food and agriculture.

## The Abuse of Women

As indicated at the beginning of this chapter, poverty has a growing feminine dimension. If stewardship in farming is to be more fully realized, then women as well as men are going to have to be enabled to understand and exercise the privileges and responsibilities of that stewardship. In many circumstances, women are handicapped by social customs and patterns that restrict their freedom and their ability to live out the biblical mandates for creation care. The inequities of the situation may be justified as biblical, but in fact nearly all lack biblical grounding. In passing, it should be noted that at least two others have seen a correspondence between the way a society treats women and the way it treats the land. This thought comes from Wendell Berry,[20] but it was developed by Walter Brueggemann[21] as follows:

> In our society, we have terribly distorted relations between men and women, [and] between *'adam* and *'adamah*, distortions that combine promiscuity and domination, precluding in both cases loyal, freely held covenantal commitments. Likely we shall not correct one of these deathly distortions unless we correct both of them. We shall not have a new land ethic until we have a new sexual ethic, free of both promiscuity and domination. Applied to the land, we shall not have fertility until we have justice toward the land and toward those who depend on the land for life, which means all the brothers and sisters.

To explore these thoughts more completely, the reader should go to the writings of Berry and Brueggemann. In this part of the chapter, I simply want to identify aspects of the abuse of women and note the biblical response to those abuses. My hope is thus to point out the ancient foundation for correcting the distortions (promiscuity and domination) that Brueggemann has named. The earlier chapters have addressed the abuse of land.

Cultures of the world, including Christian cultures, have regularly abused and devalued women. Although the revolutionary aspect of Christianity is primarily spiritual, many cultural maladies are correctable with the application of biblical teachings. At least ten categories of abuse against women are addressed in the Bible.

*1. Treatment as chattel.* Women have been treated as the property of their fathers, brothers, and husbands at various places and

times in history, with the female social function limited to reproduction, providing pleasure for men, housekeeping, and gardening.
Among other practices, daughters may be sold into prostitution.
However, the Bible teaches that marriage is not the only life option for women (1 Cor. 7:8, 34, 38; see also points 8 and 9 that
follow), prostitution is banned (1 Cor. 6:12–20), and children are
not to be sold into slavery (Neh. 5:5, 9). Wives are to be loved as
Christ loves the church (Eph. 5:25), and sometimes having no
children is counted as a blessing (Gal. 4:27).

2. *Murder.* Lethal exposure of unwanted female babies and
"honor killings" are still allowed in some cultures. The Bible pro-
hibits all kinds of murder (Exod. 20:13; Deut. 5:17; Matt. 5:21; 1
Peter 4:15). Revenge also is forbidden (Rom. 12:19).

3. *Divorce and the threat of divorce.* When the economic balance
favors men and especially when customs and laws discriminate against
divorced women, this is a powerful social and economic handicap.
Biblical teachings greatly restrict divorce for both men and women
(Matt. 19:3–9; Mark 10:2–12; Luke 16:18; 1 Cor. 7:10–16), so that
the resulting handicaps to women and children are reduced.

4. *Pornography, promiscuity, and polygamy.* These practices reduce respect for women and increase health risks to society in
general. Sexual immorality is prohibited for Christian believers
(Matt. 15:19–20; Rom. 13:13; 1 Cor. 6:9, 18; Heb. 13:4); monogamy
is exemplary (1 Cor. 7:2–3; 1 Tim. 3:2, 12, 5:9; Titus 1:6).

5. *Physical and verbal abuse.* Women sometimes face uncontrolled masculine anger and deliberate force aimed at them and
their children. Love, respect, and mutual submission replace forced
obedience in biblical teachings (Eph. 5:21, 25, 28; Col. 3:19).
Even though most husbands could physically dominate their wives,
they are urged to understand and honor them instead (1 Pet.
3:7). Scripture says we should control our anger rather than our
spouses (Prov. 29:11; Gal. 5:19–21; James 1:19–20).

6. *Physical mutilation.* Some cultures have used surgical procedures (including female circumcision) and treatments (including foot binding) for the "beautification" or control of women. In
contrast, the Bible teaches that there is a new culture (2 Cor.
5:17), that true circumcision is spiritual (Rom. 2:29), that real
beauty is on the inside (1 Pet. 3:3–4), and that severe treatment
of the body is of limited value (Col. 2:23).

7. *Prejudice and blame.* Some cultures regard women as more sinful than men, and many "Christian" cultures have ascribed the blame for the original sin to Eve and thus to all women. The New Testament redefines the cause of sin: The woman (Eve) was deceived and disobeyed God, and she had to answer to God (Gen. 3:6, 13; 1 Tim. 2:14). But the man (Adam) also disobeyed and had to answer to God (Gen. 3:6, 9–12,17). The Bible clearly places fault with him (Rom. 5:12–14). According to Christian teaching, the source of sin is not gender but the human heart (Matt. 15:18–19), and the heart needs to be changed (Rom. 2:5).

8. *Treatment as intellectual inferiors.* When education is restricted for women, illiteracy is a common result. However, in the Bible, Wisdom, both teacher and lesson, is personified as female (Prov. 8:1, 10), and feminine intelligence is praised (Prov. 31:16, 26). Women receive instruction (Luke 10:39; 1 Cor. 14:35; 1 Tim 2:11) and teach (Acts 18:26; Titus 2:3–5). This was revolutionary at the time those parts of the New Testament were written.

9. *Denial of responsibilities and rights.* Responsibilities and rights are needed for participation in stewardship, politics, inheritance, and business. Many culturally related customs and laws effectively limit what women can own and do. Rights and responsibilities are granted to both women and men for biblical stewardship (Gen. 1:28). Equity is implied for inheritance (Num. 27:8), politics (Jud. 4:4–5; Deborah was a judge), and business (Prov. 31:16). Women are individually accountable to God according to the New Testament (Rom. 14:11–12). Equal pay for equal work is not addressed by the Bible, but it is implied by the clear call for justice (Micah 6:8; Matt. 12:18, 23:23).

10. *Treatment as spiritual inferiors.* Customs and certain biblical interpretations are used to limit female participation in religious activities. However, if the Bible is accepted as the "rule of faith and practice," then Christians should agree that women, as well as men, are created in God's image (Gen. 1:27), that God chose women as the first witnesses of the Resurrection of Jesus (Matt. 28:1, 9; Mark 16:1, 9; Luke 24:10; John 20:1, 11), that women prophecy and pray in public (Acts 21:9; 1 Cor. 11:5), teach (Acts 18:26; Titus 2:3–5), and serve the churches (Acts 9:36, 16:14–15; Rom. 16:1–16). All believers are members of a royal priesthood (1 Pet. 2:9), so that there is no male or female in Christ (Gal.

3:28). The Bible leaves room for unique denominational interpretations and applications, but it is clear that a Christian perspective gives great liberty for the religious participation of women.

Whether one accepts the biblical teachings or not, it is clear that the ancient wisdom recorded in the Bible recognized and addressed many key aspects of the abuse of the poor and of women. Similar principles can be found for other groups of abused people, including children, the handicapped, and immigrants. These kinds of abuse taken together are complex and interrelated parts of the problem of increasing agricultural sustainability on a worldwide basis. Each of the twenty problems of poverty and ten problems of gender enumerated in this chapter needs some degree of resolution to improve the human condition in general and the state of agriculture in particular. One of the commendable attributes of modern Christian service is the work being done to foster economic and social justice in all facets of society, including agriculture. Examples, among many, include Heifer International,[22] International Catholic Rural Association,[23] Lutheran World Relief,[24] and the Mennonite Central Committee.[25] In the Muslim world, Nazir records other significant discussions and changes that are occurring.[26]

Because justice and equity are important aspects of agricultural sustainability, the lists constructed in this chapter are valuable because they provide particular checkpoints for justice and equity in our evaluations of any food system or of any of its parts. Hopefully identifying problems and asking questions will provide a firm footing to practically address these critical issues in the context of specific situations.

## What Is Essential Here?

This chapter has examined two aspects of human abuse. From the larger perspective of the essentials of agriculture (Appendix 3), the point is that satisfying long-term environmental and economic requirements is not enough to achieve agricultural sustainability. Agriculture is holistic, and its basic purpose is to foster human nutrition and health. Indeed, many facets of the agricultural enterprise are environmental and economic, but not all of them.

Our beginning thesis was that increasing sustainability means decreasing abuse. Agricultural sustainability requires that the most vulnerable people in the food system be recognized and treated fairly. What it means to be treated fairly goes back to our religious and ethical views, to our faith. Thus this analysis has amplified two aspects of the essentials of agriculture that were identified in previous chapters:

- There is a religious and an ethical component of agriculture.

- To be ethically sustainable, there must be a way for the disadvantaged to be fed and to recover.

It should be very clear that recovery involves more than economics, especially with regard to attitudes and treatment of women and the poor.

These concepts support a refinement of the general understanding of what it means to be sustainable. The following chapters further develop these concepts.

# Chapter Eleven

# Food, Starvation, Obesity, and Diet

## The Idealism of Youth

An environmental awakening in the United States took root in the 1960s when I was a college student already committed to a curriculum in applied ecology. Although I grew up in a setting permeated with ecological values, before the "sixties" it never occurred to me to pursue a career primarily on the basis of caring for the earth. After I outgrew my very early dreams of "being a cowboy like my dad," I became convinced that I needed to do something with my life that would provide professional and economic security. For several years I harbored the common dream of meeting those goals by one day becoming a medical doctor. Life on the ranch provided regular veterinary experience, and I imagined it was preparing me to eventually treat humans. It was a life-changing shock at sixteen years of age to discover that I responded to human blood differently than to the blood of livestock. Human blood made me sick to my stomach. I did not try to get over it. I simply changed my plans. I would become a "plant doctor," working with sap rather than blood. That decision propelled me into the field of agronomy and crop ecology.

Although my initial motivational value in choosing a career path was primarily economic and selfish, I also imagined with the idealism of youth that I could do something that would make the world a better place. The initial groundwork of the Green Revolution in crop production was being laid down by leaders in my chosen profession as I studied at the University of Nebraska, and there was much concern and discussion about reducing the threat of famine and the experience of starvation on Planet Earth. The Universal Declaration of Human Rights in 1948 had recognized the right of people to adequate food. Since that time, agronomists were caught up in the very significant contributions they could make toward meeting those internationally sanctioned goals. The 1960s and 1970s saw major breakthroughs in agricultural production. Those were exciting and heady days!

But something happened to me on my way to a career. I was startled by the realization that food production capacity did not prevent starvation, it only determined the human population at which starvation would occur. Human nature and human biology are such that more food usually results in more people, and more people not only eat more food, they also damage the natural resources needed to produce food. The best human ideas to solve the population problem usually come with so much coercion and inequity that those alternatives also are unattractive, to say the least. Changing agriculture would not be enough to solve our problem. Human behavior would have to change too. Our problem goes beyond ecology and economics and is grounded in basic human values and motivation.

That said, the problem is still with us today. It is multifaceted and full of apparent contradictions. Food production and human reproduction occupy one end of the spectrum of relevant disciplines. At the other end are the social and political components that must direct technological knowledge into effective programs that address the human rights and responsibilities for adequate food. The prism for this spectrum of knowledge and action is the desire to do what is right for others. There is still a need for idealism and hope in the face of hard reality.

The hard reality around the right to adequate food is full of such troubling contradictions! Obesity is a serious and growing health concern among the wealthy in all nations and even the poor in rich nations such as the United States and Canada. At the same time, FAO information[1] shows that at the beginning of the twenty-first century, about 25,000 people die each day of starvation (over 9 million each year), and about 800 million people are malnourished, including 75 percent of the world's farmers. Although indirect, there is a connection between what we eat and food supply for the rich and poor alike. Current headlines and ancient wisdom both address aspects of the problem. The goal of this chapter is to consider those relationships and the biblical guidelines for diet in light of the two-edged food crisis of obesity and starvation that troubles us in the modern world.

## Hunger and Starvation

Famine is a sign of human mismanagement of the food supply. According to the Bible, it also represents God's judgment for such human failings. Famine is mentioned ninety-one times from the first book of the Bible (Gen. 12:10) to the last (Rev. 18:8), an indication that it is a persistent problem. The story of Joseph in Genesis 41–47 is, among other things, a story of how planning and grain storage in times of surplus can alleviate the starvation associated with a prolonged famine. The early books of the Bible seem to regard famine as part of a natural cycle for which God provides wisdom and blessing, so that survival and prosperity are possible in an insecure world. Later, in the prophetic books, famine is listed as one of the means of God's punishment for the nations of ancient Israel and Judah that have broken their relationship with him, with one another, and with the land. Starting in Jeremiah 5:12, the prophet links the sword and famine twenty-three times as vehicles of God's judgment. One of the themes of Jeremiah is that God's judgment comes because people have ceased to care for the land (Jer. 2:7, 9:10–13, 12:10–13), so there is an association of land care and famine. There also is an association of famine with war, a relationship that is still true today. The New Testament writers acknowledged both the natural and judgmental aspects of famine (Luke 15:14; Rom. 8:35; Rev. 6:8, 18:8) and reported that famines occurred at the time the New Testament was being written (Acts 11:28).

In the face of human hunger, the Bible teaches that we are to share our food. The story of the seven-year famine in Genesis shows that Joseph sold food to the hungry from other nations. The farmer or merchant who sells food instead of hoarding it during hungry times will be blessed according to Proverbs 11:26. In time of war, the prophet Elisha counseled the following for those held as prisoners:

> . . . Would you strike down those whom you have taken captive with your sword and with your bow? Set bread and water before them, that they may eat and drink and go to their master." (2 Kings 6:22b, ESV)

Citing the book of Proverbs, Paul similarly advised feeding those who regard us as enemies:

> [20] . . . [I]f your enemy is hungry, feed him; if he is thirsty, give him something to drink; . . . [21]Do not be overcome by evil, but overcome evil with good. (Rom. 12:20b–21, ESV)

If our enemies are to be so treated, then we also should share our food with everyone who is hungry and  within our reach. Jesus taught as follows:

> [35][Referring to a king's judgment, Jesus said,] "For I was hungry and you gave me food, I was thirsty and you gave me drink, I was a stranger and you welcomed me." [37]Then the righteous will answer him, saying, "Lord, when did we see you hungry and feed you, or thirsty and give you drink?" [40]And the King will answer them, "Truly, I say to you, as you did it to one of the least of these my brothers, you did it to me." (Matt. 25:35–40, ESV)

The Christian attitude toward feeding the hungry is partly shaped by the New Testament accounts of the miracles performed by Jesus in feeding thousands of people (see Matt. 14:13–21, 15:32–39; Mark 6:30–44, 8:1–10; Luke 9:10–17; John 6:1–15):

> Then Jesus called his disciples to him and said, "I have compassion on the crowd because they have been with me now three days and have nothing to eat. *And I am unwilling to send them away hungry, lest they faint on the way.*" (Matt. 15:32, ESV; emphasis added)

Feeding the hungry when we have an abundance of food is easier than when food is in short supply, as during a famine. However, the "Golden Rule" of Jesus (which has a parallel teaching in just about every religion of the world) bids us to treat others the way we would want to be treated ourselves (Matt. 7:12; Luke 6:31). For the committed person, caring for the hungry is a matter of pleasing God. The Christian writer of the book of Hebrews stated the law of compassion as follows:

> Do not neglect to do good and to share what you have, for such sacrifices are pleasing to God. (Hebrews 13:16, ESV)

The apostle James noted the negative consequence of not showing compassion:

> So whoever knows the right thing to do and fails to do it, for him it is sin. (James 4:17, ESV)

With this perspective, feeding the starving is a primary motivation of many Christian mission organizations. In addition to simply bringing food to the places of great need, there also is an effort by some groups to help local people develop their own food production resources so that famine is less of a threat for them.[2, 3] Groups such as Educational Concerns for Hunger Organization (ECHO), the Mennonite Central Committee, and Mission Moving Mountains provide good examples. R. L. Wixom lists other similar groups.[4] Many of the world's religions share the theme of compassion, and the model has spread as well to governmental agencies representing many countries and to many secular nongovernmental organizations (NGOs) representing private donors and philanthropic foundations. L. Shannon Jung's book *Sharing Food*[5] provides a recent Christian perspective on the related issues.

## Obesity

Some cultures regard obesity as a sign of wealth and blessing. In modern Western culture, the slim, athletic model is idolized, and obesity is recognized as an epidemic health risk.[6] The scriptures of the Hebrew Bible treat obesity as a sign of self-sufficient complacency that soon leads to spiritual trouble:

> [God said,] For when I have brought them into the land flowing with milk and honey, . . . and they have eaten and are full and grown fat, they will turn to other gods and serve them, and despise me and break my covenant. (Deut. 31:20, ESV; see also Deut. 32:15, Neh. 9:25)

Psalm 73 refers to "eyes [that] swell out through fatness" in conjunction with "hearts [that] overflow with follies" (verse 7, ESV). The prophet Jeremiah also links fatness with a self-centeredness that is evil:

> [T]hey have grown fat and sleek. They know no bounds in deeds of evil; they judge not with justice the cause of the fatherless, to make it prosper, and they do not defend the rights of the needy. (Jer. 5:28, ESV)

The New Testament continues the argument that obesity is not associated with the right way to live:

> You have lived on the earth in luxury and in self-indulgence. You have fattened your hearts in a day of slaughter [which is the day of God's judgment on behalf of the abused]. (James 5:5, ESV)

The biblical references though are somewhat symbolic, focusing more on the condition of the heart (the inner soul) than on the condition of the body. But the ancient wisdom is clear in asserting that there is a relationship between body and soul and between the physical and the spiritual. Agur, the sage, cried out as follows:

> [8]Remove far from me falsehood and lying; *give me neither poverty nor riches; feed me with the food that is needful for me,* [9]lest I be full and deny you and say, "Who is the LORD?" or lest I be poor and steal and profane the name of my God. (Prov. 30:8–9, ESV, emphasis added)

Thus too little and too much to eat both lead to undesirable temptations.

In contrast, David, the psalmist, saw the occasional celebration with rich food as consistent with praising God:

> My soul will be satisfied as with fat and rich food, and
> my mouth will praise you with joyful lips. (Ps. 63:5, ESV)

In Psalm 36:8, the same author spoke of the "feast of abundance" in the house of God. David's son, Solomon, indulged in all pleasures, including feasting (Eccles. 2:10), and he concluded that a legitimate purpose of feasting was strength:

> Happy are you, O land, when your king is the son of the
> nobility, and your princes feast at the proper time, for
> strength, and not for drunkenness! (Eccles. 10:17, ESV)

The book of Esther also explains feasting as a part of the celebration of God's goodness to his people (chapter 9), and the prophet Isaiah describes feasting as a part of God's future kingdom (Isa. 25:6). Elsewhere, it is called "the marriage supper of the Lamb" (Rev. 19:9, ESV).

The modern problem of obesity is thus recognized in the ancient writings as a potential difficulty because of the pleasures associated with eating and the adverse bent of human nature. However, the extent of the problem can only be understood in detail with a modern analysis. In the United States and Canada, the food industry is trying to sell approximately twice as many calories as the population needs. The whole goal of advertising is to get us to eat all they have to sell. Often under the guise that this food is good for your health, young and old alike are prompted to eat more. The fact is, more and more people are giving in to advertising pressure.[7] The problem also is associated with "cheap food" and the great success of the fast-food industry. Fast food has already had serious environmental and social consequences, and it contributes to caloric intake that makes us fat.[8] One of the great misconceptions is that inexpensive food is good for the poor. The cheapest food also is the poorest food in nutritional quality, loaded with fat and salt. Thus the obesity epidemic is rampant among the poor, who cannot afford a better diet. In addition, the industrial patterns that reduce the cost of preparing and distributing food also eliminate many jobs. The result is that "cheap food" contributes to keeping the poor in their poor state while decreasing their health and increasing their weight. The problem is complicated and will not be easily solved. It is more likely to be solved when ethical principles and religious values are applied.

Returning to the ancient scriptures, we are advised to control our eating:

> [2]If you are a big eater, put a knife to your throat [i.e., control yourself], [3]and don't desire all the delicacies— deception may be involved. (Prov. 23:2–3, NLT)

(It almost sounds as though the ancient writer knew about modern advertising.) The New Testament counsels believers to take care of their bodies, because they are the home of the Holy Spirit (1 Cor. 6:19). The biblical standard for eating, as for all things, is in relationship to God:

> So, whether you eat or drink, or whatever you do, do all to the glory of God. (1 Cor. 10:31, ESV)

## The Biblical Diet

With current problems of obesity, there also is great interest in dieting. Some diets are being designed for believers and claim to be biblically based. Examples can be found on the Internet with both positive and negative analyses. The goal here is not to identify and evaluate the possibilities but to note the main themes that run through Scripture as they are related to the way we eat and their relationship to agricultural sustainability.

*Vegetarianism* appears to be the first mode of eating described in the Bible. In the very first chapter, God announced that he had given the human race the "seed-bearing plants" and "all the fruit trees for your food" (Gen. 1:29, NLT). However, just a few chapters later, a new order is established, in which "All the wild animals, large and small, and all the birds and fish" were given as food for humans, just as God had earlier given them "grain and vegetables" (Gen. 9:2–3, NLT). Since livestock were clearly a part of the agricultural system from a much earlier time (Gen. 4:4), they may have been a part of the diet even before the wild animals were approved as food in Genesis 9.

Reference to a vegetarian diet did not disappear with the passage of time. Daniel and his companions thrived on a diet of vegetables (Dan. 1:12–15), and the apostle Paul refers to followers of Jesus Christ who ate only vegetables (Rom. 14:2). However, the main dietary theme of the New Testament is on the elimination of dietary restrictions. For example, Jesus reinterpreted the Hebrew concept of diet and defilement (becoming unclean):

> [15]There is nothing outside a person that by going into him can defile him, but the things that come out of a person are what defile him . . . [19]since it [all food] enters not his heart but his stomach, and is expelled. (Thus he declared all foods clean.) (Mark 7:15, 19, ESV)

A dietary lesson also was given to the apostle Peter (Acts 10:11–16) to show both that God had included all kinds of food and all kinds of people among those things that are acceptable to him. The Scripture also warns that there will be false teachers who advocate a restricted diet as a part of their religious system (1 Tim. 4:3). But on the contrary, the true believer will argue as follows:

> [4]For everything created by God is good, and nothing is to be rejected if it is received with thanksgiving, [5]for it is made holy by the word of God and prayer. (1 Tim. 4:4–5, ESV)

Vegetarianism is today appealing to those who wish to make a statement against the ecological and social damage being done by a grain-based and industrialized meat production system. Ecologically and socially sound methods of meat production are available (especially "pasture-based systems," see Chapter 7), and our own research indicates that sustainable land management systems in areas with some hilly topography will feed the most people when meat and dairy products are included in the diet. In fact, vegetarian systems that meet caloric needs with crops such as soybeans can damage the environment, because such crops do not provide the amount of soil protection offered by perennial forages used as livestock feed. Hopefully our society will be able to transition from present damaging meat production methods to pasture-based systems that are socially, ecologically, and nutritionally superior. Fattened calves and poultry are parts of the biblical feast (1 Kings 4:23, KJV, Luke 15:23, and elsewhere), but leaner, pasture-finished beef also is mentioned (1 Kings 4:23), and modern science is showing that grass-fed animal products, though they presently cost more at the market, are both better for our health and for the natural and social environment.[9]

With regard to the present dilemma related to eating grain-fed industrialized meat, Paul's discussion of vegetarianism in the book of Romans (14:1–6) may be doubly relevant. There the main concern was if a believer could eat meat sacrificed to idols. In a similar vein, today's problem is if one concerned about agricultural sustainability can eat meat produced for the "economic idols" of this age. Paul says that "the weak person eats only vegetables" (Rom. 14:2). The weakness is not in body but in faith (Rom. 14:1). Paul's conclusion is that believers should not judge one another in the matter of diet. Vegetarianism is alright, and meat eating is alright. Diet selection today continues to be a matter of faith. Some devout believers are trying to change a faulty meat production system, either by not eating meat or by eating only that meat produced in a more sustainable manner. Paul states the bottom line on the Christian diet as follows:

> [20]Don't tear apart the work of God over what you eat. Remember, there is nothing wrong with these things in themselves. But it is wrong to eat anything if it makes another person stumble. [21]Don't eat meat or drink wine or do anything else if it might cause another Christian to stumble. (Rom. 14:20–21, NLT)

Christians usually think of this kind of stumbling in terms of loss of faith, but there also is loss of livelihood, or health, or a sustainable future.

A friend once offered the opinion that Paul's advocacy of a vegetarian diet under some circumstances provides an example of biblical relativism. I prefer the interpretation that it provides an example of adaptation to varying circumstances. The underlying principle is important for agricultural sustainability, because one aspect of sustainability is adaptability. A sustainable system will have options that allow adjustment to changing circumstances. The apostle Paul applied the principle of adaptability to diet because he was concerned about people. Modern nutritional guidelines also are concerned about people, though in a different way, and they likewise include options for providing a balanced diet for people with different health and cultural needs.

*The USDA Food Pyramid* (MyPyramid.gov[10]) is an example of modern dietary guidelines designed to meet nutritional needs. The 2005 pyramid is composed of a set of connected triangles, and the size of each triangle roughly approximates how much of six food groups should be consumed. An illustration of the vegetarian option for diet is in the meat-and-bean triangle, which includes both as a means of meeting the dietary protein requirements. Grain-based foods are in the largest triangle and are supposed to be consumed in the largest quantities. In the Bible, grains are repeatedly recognized as the foundation of diet (Gen. 9:3, 41:49; Acts 27:33–37; Rev. 6:6). Psalm 104:15 refers to "bread to give them [the people] strength." Though there are contrary ideas about the benefit of low carbohydrate diets, ancient wisdom and the mainstream of modern nutritional knowledge agree that grains are important.

"Oils and fats" occupy the smallest triangle of the pyramid and are to be consumed "sparingly." According to kosher food laws, the fat of meat (along with the blood) was not to be consumed at all (Lev. 3:17). Olive oil (the main biblical oil) was used for fuel for lamps (Exod. 27:20; Lev. 24:2; Matt. 25:3), for anointing or soothing the skin (Ps. 23:5, 133:2; Isa. 1:6; Ezek. 16:9), for medication (Luke 10:34), for offerings (Lev. 7:12; Num. 7:13; Ezek. 45:24), and for cooking and food (Lev. 6:21; Num. 11:8; Deut. 14:23; 1 Kings 17:12). The amount of oil actually consumed in food is not clear, but the ancient Hebrew concern about obesity has already been noted. The 2005 pyramid also identifies "discretionary calories" that come from treats and snacks rich in fat and sugar. They also are to be consumed sparingly. Honey is the term used for the main sweeteners of Bible times, especially wild bees' honey and the sweet syrup made from dates. The goodness of honey was recognized (Prov. 16:24, 24:13), but one was not to eat too much of it (Prov. 25:16, 27). Thus today's nutritional recommendations and the Bible agree about what foods should be limited.

The food pyramid also includes four other groups—fruits, vegetables, milk, and meat and beans. As already explained, a variety of optional foods is included in each group. The biblical record does not indicate how much of each group should be eaten, but all are mentioned as part of the diet. Meat has already been discussed. Fruits, and especially grapes, are frequently mentioned as food (e.g., Lev. 25:3; 2 Sam. 16:1; Jer. 40:12; Matt. 7:16). Vegetables also are noted as a part of the diet (Deut. 11:10; Dan. 1:12, Rom. 14:2). Grain legumes (beans, lentils, dried peas, etc.) are mentioned in Genesis 25:34 (lentil stew) and Ezekiel 4:9 (mixed in bread), among other places. Dairy products mentioned in the Bible include milk, cheese, and curds (see Chapter 7). Eggs from wild and domestic birds also are noted (Deut. 22:6; Job 6:6 (KJV), 39:14; Isa. 10:14; Luke 11:12). Some dietary guidelines also include wine because of its reputed health benefits when consumed in moderation.[11] In one place, the Bible also advises drinking a little wine (1 Tim. 5:23), but in general it warns against excessive drinking and addiction (Prov. 23:31; Eph. 5:18; 1 Tim. 3:2–3, 8; Titus 1:7). It says that liquor is for the dying and the depressed (Prov. 31:6), and that believers should be careful not to lead others into bad habits by their own consumption of alcohol (Rom. 14:21).

Row crops such as corn (maize) and soybeans, so important in grain-fed meat production, were unknown in the ancient Near East. Both are productive on the best crop land, but they often are planted at relatively wide spacings that do not adequately protect the soil from accelerated erosion on sloping lands. In addition, soybeans return relatively little organic matter to the soil. Thus it would be good for the environment, and especially for soil health, if row crops could be partially replaced with sod-forming perennials that better protect the soil. Sod-forming perennials are mostly forages suitable as feed for pasture-raised livestock. When animal production is based on rations high in pasture, hay, silage, and food by-products, then the food supply is increased by using livestock to convert those feeds that people do not eat into meat and milk and other animal products that are suitable for humans. Such systems can benefit the environment as well.

We also need to be aware of the efficiency of livestock production systems in converting energy and protein in the animal diet to energy and protein in food that humans can eat. That efficiency is increased when animals are allowed to live longer lives. That mainly happens by eating dairy products and eggs instead of meat and poultry, because with dairy and eggs, the producing animal does not have to be killed in order to get the food. Life expectancy is further increased by letting the animals live a more natural life at lower production levels on pastures instead of maximized production rates on factory farms.

## A Different Model of Food Production

All of this points to changes in the way our modern food production systems are designed. Two kinds of agricultural systems are evolving in the world today. One is based on an industrial model characterized by fewer farmers, bigger farms, and increasing production through the application of the latest scientific and industrial methods. This is sometimes called the "conventional" model. The other is an "alternative model," still somewhat poorly defined, but focused on change based on sustainability, holism, and the health of farm communities. A major argument in favor of the dominant conventional model is that we have such a high and growing demand for food worldwide that we must make agriculture as productive as possible. It is certainly true that when we maximize production per acre or hectare, less land is needed for agriculture and more land may be left in its wild state for the rest of the natural world. In addition, the conventional model keeps food costs as low as possible, and it supports a globalized economy that many believe offers the greatest freedom and economic equity for the most people. The proponents of the alternative model argue that the conventional system is not sustainable. For one thing, it is too dependent on external energy from fossil fuels. It also results in fewer farmers with consequent loss of rural communities. In addition, it has questionable effects on ecological and human health. There must be a better way.

One step that has already been taken is to design socially conscious diets. These are represented by cookbooks such as *Diet for a Small Planet,*[12] and *More-with-Less Cookbook.*[13] They offer eating alternatives that can be applied today to help resolve the dietary dilemma. Mark Graham's book on sustainable agriculture[14] develops an ethical argument for modifying our food choices and food production methods. Studies also are underway that assess the relative productivity and social costs of the competing agricultural systems. The reader is encouraged to watch and evaluate the growing pool of comparative information as it becomes available. Looking to the future, various technologies are being marshaled to attack the problem of providing a healthful, equitable diet. Our own research with computer models is an example of the kinds of tools being developed to help decision makers chart the path to the future of our food system. However, a wise mentor once told me to be cautious about models. They can be used in such a way that they are no more than self-fulfilling prophecies. What we first need to do is discern the kind of world we want for ourselves and future generations. Then models can help us ascertain if our goals are compatible and how we might reach them. Similar guidelines can be applied to all kinds of technologies.

## Change

The ancient writers we have been studying envisioned a world with adequate food and a varied diet, time for celebration, and a set of relationships with one another, with nature, and with their God that provided meaning and expressive joy. Our challenge for change is so large that we need the best of past and present insight, and we need it from the spiritual as well as the material perspectives on life. The prophet Jeremiah was inspired to say that we need "plans for wholeness and not evil, to give . . . a future and a hope" (Jer. 29:11, ESV). My youthful idealism has been dented and tarnished by life's hard realities, but I still see a reason to hope for a better food system. The reason is that adding a spiritual context provides motivation for people to change. The lens of faith will allow believers to clearly see the unselfish principles that are the foundation of sustainability. On a worldwide basis, most farmers and consumers are still people of faith. Thus when the positive elements of faith are integrated into education about food and farming, there is hope for a better world.

# Chapter Twelve

## Agricultural Sustainability

If we understand something, we should be able to define what it is. However, the concept of agricultural sustainability seems to be especially complex. In the preceding chapters, fifteen essentials of agriculture have been identified (Appendix 3). Then the ethical aspects of the subject were expanded by identifying twenty causes of poverty and ten abuses of women that affect agriculture. Finally, a focus on food and diet highlighted current weaknesses and unsustainable aspects of the modern food system. In fact, each of the eleven preceding chapters presents a unique perspective that could lead to its own definition of agricultural sustainability. And such definitions exist. There also are multiple related concepts, such as agroecology, alternative agriculture, biodynamic agriculture, ecological agriculture, holistic agriculture, low-external input agriculture, natural farming, permaculture, regenerative agriculture, sustainable agriculture, and the whole food movement. One might say that this topic is notoriously difficult to define to everyone's satisfaction.

Nevertheless, if we are to convince others that this is a subject worthy of earnest and prolonged attention, then we need to be able to define it in a simple manner and to relate it to the serious concerns of all thinking people. Thus in this last chapter, my goal is to move from complexity to simplicity and in so doing to tie all of the parts together and make them as understandable as possible.

## Complex Concepts of Agricultural Sustainability

Researchers Beus and Dunlap[1] conducted a comprehensive analysis of the literature and identified six differential elements that defined agricultural sustainability, or what they called "alternative" agriculture. Their lists included (1) relative independence from industrial and technological inputs, (2) decentralized or more local production and management, (3) community as opposed to a competitive orientation, (4) harmony with nature instead of control, (5) diversity instead of specialization of enterprises on individual farms, and (6) restraint reflected through consideration of environmental and social costs and cautious application of new technologies. (This last element often is called the "precautionary principle"[2]) Their concepts combine many of the points made in the preceding chapters, and are particularly focused on the social or cultural aspects of agricultural sustainability in Chapter 9.

In a subsequent study, Chiappe and Flora[3] noted that no women were involved in the earlier study.[4] They proceeded to interview twenty-five women practicing agricultural sustainability in Minnesota, and their data confirmed the six elements identified by Beus and Dunlap, though with nuances of feminine emphasis. In addition, they found two other factors that characterize a sustainable approach that the first study had not noted. Continuing the first list, they also stressed the importance of (7) quality of family life and (8) spirituality as motivation for the way they approach farming.

It is interesting that there is apparently a gendered aspect of recognizing the spiritual components of agricultural sustainability. Certainly men have written on the topic. Many of the writings of Wendell Berry (cited elsewhere) have clear themes related to values or religion. Fellow agronomist Roger Elmore[5] wrote an article about the interplay of religious worldviews and the processes of agricultural sustainability. He concluded that the moral or religious aspect of agriculture is very important but largely ignored in many of the secular models of agricultural development. This is confirmation that the spiritual component often has not been acknowledged.

However, the situation is changing, in part because of contributions such as those mentioned earlier. A recent multiauthored book, *Biblical Holism and Agriculture*,[6] advocates a Christian approach to agricultural sustainability and emphasizes the application of scriptural principles to a wide range of topics related to agriculture. The bulk of that work focuses on the social and ethical dimensions that go well beyond what I have presented in these chapters. *Faith in Conservation* published by the World Bank[7] in 2003, and *Inspiring Progress*, by Gary Gardner,[8] published in 2006, add to the discussion the religious perspective of several faiths.

With this background, it is interesting to look at an "official" definition of sustainable agriculture used by the U.S. government (see boxed text that follows). The ethical or religious aspects are there if you look for them, but they are not explicit. The usual analysis of this definition recognizes three main components of sustainable agriculture: biological-environmental (points 1–3), economic (points 2 and 4), and social (point 5). The core argument is that for agriculture to be sustainable, it must satisfy ecological, economic, and social criteria.

---

### Definition of "Sustainable Agriculture" Used by the U.S. Government

Sustainable agriculture is

". . . an integrated system of plant and animal production practices having a site-specific application that will, over the *long term*:

1. satisfy *human food* and fiber needs;

2. enhance *environmental quality* and the *natural resource base* upon which the agricultural *economy* depends;

3. make the most efficient use of *nonrenewable resources* and on-farm resources and integrate, where appropriate, *natural biological cycles* and controls;

4. sustain the *economic viability* of farm operations; and

5. enhance the *quality of life* for farmers and *society as a whole*."

*Source:* M. V. Gold, "Sustainable Agriculture: Definitions and Terms," *USDA Special Reference Brief 94-05* (Beltsville, MD: National Agriculture Library, 1994), emphasis added.

It has been difficult to agree on definitions, in part because the social and economic elements are so relative, so dependent on current economic circumstances and value systems that vary from culture to culture. This frustrates ecologists who would like to have more stable standards that would enable them to rigorously define concepts such as "soil health" and "wise use." Thus T. E. Crews and colleagues defined sustainable agriculture solely in terms of ecological principles.[9] However, this will not do, at least for those who want to take a larger view that includes the social and economic alternatives of the way we do agriculture. Food and agriculture are about more than ecology. Thus a systematic thinker such as David Orr argued that agriculture is one of the liberal arts and needs to be an integrated part of the requirements for a liberal education.[10] The two sides of the debate are examples of the arguments of agriculture explained in an excellent little book of that name by Jan Wojcik.[11] Wojcik sees that most of agriculture's controversies can be characterized by two perspectives. One is made up of the "progressives" who generally favor free markets and the latest scientific technologies. The other is made up of the "sustainers" who are more precautionary and take a view based on traditional practices and more holistic ecological principles. Both sides include economic and social elements as well as biological factors in their understanding of agriculture.

However, I am not quite satisfied with the ecological, economic, and social triad. There are two concerns. The first is that economic and social elements are not really distinct. The economic system is a component of the larger cultural aspect of human society. One might argue that the economic factor should be singled out because it represents a leverage point in the system, a place where relatively small and managed changes can have big results. A counterargument is that too much emphasis on economics can be used to justify the false assertion that if something is profitable, it is sustainable. To be truly sustainable, ecological and other social factors must also be favorable for the long term, as the official definition maintains (see boxed text on sustainable agriculture).

The second concern is that the religious or ethical aspects are largely hidden when ecological, economic, and social elements are identified as the primary components. My preference is to combine the economic and social elements and add a new primary factor that covers the religious and ethical aspects of our subject. That aligns both with the growing emphasis on the spiritual dimension and with the traditional division of academic knowledge into the biophysical sciences, the social sciences, and the humanities.[12]

It is also worthwhile to ponder what is being defined. Is *sustainable agriculture* the best terminology? At first it seems that *agriculture* should be the subject, but there is a problem with this. The implication is that there is a kind of agriculture that has reached a desirable end point, that there is a kind of agriculture we can be sure is sustainable. My experience and observations are less confident than that, so my preference is to reverse the word order to *agricultural sustainability*. By emphasizing *sustainability*, the reality of an unfolding and a future-oriented subject is more accurately represented. With the foregoing analysis in mind, a revised presentation of the component topics of agricultural sustainability is offered in the boxed text that follows.

---

### A Word Model of the Essentials of Agricultural Sustainability

Progressive improvement in agricultural sustainability is based on sound and nondestructive

A. *Ecological relationships* that conserve or enhance

1. biodiversity, especially of crop and livestock resources and the wild species associated with food production and human health.

2. land and climate resources.

3. soil and water resources and our knowledge of their management.

B. *Social relationships* that foster

1. economic processes and policies that are fair and effective in favoring best management practices.

2. an infrastructure that allows farmers to get the food they produce to consumers at the time and place they need it and to receive from the rest of society the goods and services they need and desire.

3. a knowledge system that discovers, refines, and transmits essential information from generation to generation.

C. *Spiritual and ethical relationships* that holistically encompass a worldview to

1. provide joy and quality of life to farmers, their families, and all who work in agriculture so that people will continue to want to farm.

2. engender stewardship or care of the natural and social foundations of the food system.

3. address injustice, economic abuse, and other social or cultural failures that endanger the continued health of any part of the food system.

---

## Metaphors of Agriculture

The model of agricultural sustainability presented here summarizes the preceding chapters. However, for the general public, a simpler summary is important. The three-legged stool offers a useful metaphor. The seat of a stool is a place where some object or concept can be supported. The three legs, if stable and of about equal size and strength, can successfully support the main subject of concern. As a teaching device, this is almost a perfect model—one main concept supported by three points.

The three-legged metaphor can be related in two ways to agricultural sustainability. The first is to take an archeological and anthropological view.[13] From those perspectives we see that civilization, with its cities and customs of higher learning, emerged from an agricultural base. Thus this metaphorical stool has a seat of agriculture that supports civilization. The identity of the three legs will depend on the perspective of the teacher. Given the viewpoint of an agronomist (a crop and soil scientist), those legs will represent the main categories of natural resources necessary for agriculture, namely, the resources of (1) biology or biodiversity, (2) land and climate, and (3) soil and water.

The biological leg of the stool includes domesticated crops and livestock as well as all kinds of living things that drive the natural cycles and flows of matter and energy in an agroecosystem (agricultural ecosystem). In addition to livestock, honey bees and nitrogen-fixing bacteria are among the farmer's helpers, and all of their close and distant biological relatives may play some role, beneficial, harmful, or neutral, from the farmer's point of view.

The second leg of the stool as a metaphor for agriculture is the land and associated climate taken together. Good land is a limited resource that is required for agricultural sustainability. The productive limits for crops and livestock are set by land and climate. Bananas cannot survive the winters in Wisconsin. English breeds of cattle do poorly in the hot and humid tropics. Corn and wheat cannot be grown where it is too dry, and few crops besides rice and taro prosper where it is very wet. Climatic moisture regimes and temperature patterns set the limits of where various kinds of food can be produced, thus explaining the very serious agricultural concerns associated with global warming.

The soil and water resources also are emphasized by agronomists and ecologists because we cannot have agriculture without fertile, tillable soils that have adequate water supplies. Many changes in the soil will be lasting ones, so soil management is a critical issue for agriculture. When water is considered with soil, they together form one of the bases for agricultural productivity.

In this first metaphor, the picture of agricultural sustainability is essentially an ecological picture with the social and spiritual components hidden as parts of civilization, which is seated on the stool. I believe the metaphor can be improved by basing it on the more holistic detail of the boxed text on a word model of the essentials of agricultural sustainability.

With this second metaphor, the seat of the stool is agricultural sustainability. Civilization still sits on the seat, but the legs are redefined to represent (1) a biological and an ecological component, (2) a social and an economic component, and (3) a spiritual and an ethical component. The overall picture is that agriculture is a human and social endeavor with natural elements. In comparison, the first metaphor represents an ecological viewpoint that agriculture is a part of the larger natural world, and that it stands on a foundation of natural resources. The second represents a viewpoint that agriculture is a part of larger human culture, and that natural resources are just one component in the agricultural system. Only the second metaphor captures the comprehensive associations of agriculture and links it to the rich cultural and religious significance of food.

Although the three-legged stool is a very useful metaphor, it suggests more stability (or less dynamism) than is true of agricultural sustainability. That brings us to the possibility of a one-legged stool. My colleague Gil Gillespie, at Cornell University, has already suggested that this might be a better model.[14] Both Gil and I milked cows by hand while we were growing up as farm boys at a time when small herds were still economical. Both of us used a one-legged stool. The construction could be as simple as two segments of 2 × 4 lumber about 12 inches (30 cm) long. They were nailed together to form a "T". Often the one leg was of larger dimension, perhaps a 4 × 4," so that the seat could be more securely attached to the leg and the leg had more of a foot on which to stand. Using such a stool to milk a cow was a bit of a balancing act, but it was normally one of the less complicated and more reliable aspects of a process that had numerous challenges related mostly to the cow.

For such milking, the cow usually was confined with a stanchion that fitted around her neck. One would sit by the udder on the right side of the cow, balancing on the one-legged stool by using one's own legs as the second and third support for one's body. With one's head against the flank of the cow, it was routine to balance holding a bucket between the legs while rhythmically stripping milk from the teats. The complications came from also trying to dodge the switching of the cow's tail or properly anticipating her surprised moves when dogs or cats or strangers showed up in unexpected places. If one paid attention to the process, it was simple enough to do. (Nevertheless, it was a wonder for most of the uninitiated when they got to see it done.)

The great thing about the one-legged stool as a metaphor for agricultural sustainability is that it captures some of the dynamics of attentiveness and balancing that are parts of agriculture. If we imagine setting the single leg on different kinds of ground, the metaphor even captures some of the place-specific good farming emphasized by Jackson[15] and Berry.[16] The difficulty of this metaphor is that it is somewhat unfamiliar and perhaps unbelievable. The difficulty can be corrected, though, because it can be easily demonstrated that sitting on a one-legged stool is possible. But the stool will not stand up by itself. That alone is a useful addition to the metaphor, because agriculture also is a human construct that will not stand up by itself. Agriculture takes constant effort and adjustment to keep it going. Without humans, natural processes would soon obliterate agriculture.

The change of metaphor to a one-legged stool has raised another question. What happened to the components separated into three legs in the first metaphor? The answer is that they are now wrapped together in the one leg. Another weakness in the first metaphor is that it implied a degree of separation that is not an accurate picture of agricultural reality. For example, cultures and social systems are very much affected by the physical environment. That is very obvious for primitive cultures, but it also is obvious in the cultural "necessities" of air conditioning in Phoenix, Arizona, and furnaces in Toronto, Ontario. And of course, air conditioners and furnaces have impacts on both the local and global environments. Likewise, economic factors, such as grain and meat prices, have ecological consequences or interactions, such as the burning of the Amazon forest and the plowing of the Great Plains. The economic and social components are also in constant interplay, and in fact, one can argue that more realism is represented when they are not separated.

Figure 1 illustrates how the components of agricultural sustainability might be linked, similar to the ecological principle "everything is connected to everything else."[17] For the sake of defining and describing the system, having three main parts aids understanding, but we must remember that they are interconnected and interdependent parts.

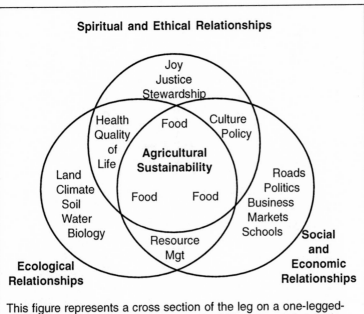

This figure represents a cross section of the leg on a one-legged-stool model of agricultural sustainability. It captures the interconnectedness of the components, but one must sit on a one-legged stool to understand the dynamic balancing implied by such a model. The list of dimensions shown in each circle is representative and not comprehensive.

Figure 1. A Venn Diagram of Agricultural Sustainability

This figure is too small to include all of the components of sustainability, but the main elements are shown in close proximity, providing more realism than is possible with three separate legs. The result is a new metaphor:

> Agricultural sustainability is like a one-legged stool supported by ecological, social-economic, and spiritual-ethical relationships interwoven into a single supporting leg. To make the stool stand, we must sit on it, thus providing with our own legs the sustainable balance of both attention to process and orientation to the future.

## Simple Facts

In the opening chapter, I argued that it is all about food and that food is really the central and comprehensive theme linking all of the details of all of these chapters. We have now returned to that place. In the metaphor of the one-legged stool, described earlier, for agricultural sustainability, it is food, or interest in food, that holds together the diverse elements in the single leg. Food is what transcends and integrates all of the aspects of agricultural sustainability. And the wonderful thing about food as a central and unifying concept is that it touches every person. We all need food as long as we live in this world.

The other wonderful thing about food as the central theme is that it gives everyone a role in the process of developing a more sustainable agriculture.[18] When we study systems, we look for leverage points. A leverage point in the food-agricultural system is the individual eater. At a recent conference, Brother David Andrews made the point this way: "The fork and the spoon are levers that shape culture."[19] Our food choices affect not only our own health but also the compensation that farmers and agricultural workers receive for their work and the kind of environment in which they must work. How we eat, and how we acquire the food we eat, sends a small message throughout the food system, which influences the direction of the whole system. Choosing to eat food produced in the most sustainable way we can find and afford will make a difference.

When you plan your next meal, think about all of the different ways food connects to life, and then eat wisely.

# Appendix 1

# Abbreviations and Citation of Books in the Bible

The following is an alphabetical list of the standard abbreviations for the books of the Bible used in the chapters in this book. Verses of the Bible are cited by giving the book (usually abbreviated), the chapter, and the verse. The chapter and verse are separated by a colon, thus Gen. 1:1 refers to the first chapter and first verse of the book of Genesis. The citation Ps. 91:11–13 refers to the book of Psalms, chapter 91, and verses 11 through 13. A list of the books of the Bible in the order in which they occur is found in Appendix 2.

| | | | |
|---|---|---|---|
| Acts | Acts | Exod. | Exodus |
| Amos | Amos | Ezek. | Ezekiel |
| 1 Chron. | 1 Chronicles | Ezra | Ezra |
| 2 Chron. | 2 Chronicles | Gal. | Galatians |
| Col. | Colossians | Gen. | Genesis |
| 1 Cor. | 1 Corinthians | Hab. | Habakkuk |
| 2 Cor. | 2 Corinthians | Hag. | Haggai |
| Dan. | Daniel | Heb. | Hebrews |
| Deut. | Deuteronomy | Hos. | Hosea |
| Eccles. | Ecclesiastes | Isa. | Isaiah |
| Eph. | Ephesians | Jas. | James |
| Esth. | Esther | Jer. | Jeremiah |

| | | | |
|---|---|---|---|
| Job | Job | Ps. | Psalms |
| Joel | Joel | Rev | Revelation |
| John | John | Rom. | Romans |
| 1 John | 1 John | Ruth | Ruth |
| 2 John | 2 John | 1 Sam. | 1 Samuel |
| 3 John | 3 John | 2 Sam. | 2 Samuel |
| Jon. | Jonah | Song Sol. | Song of |
| Josh. | Joshua | | Solomon |
| Jude | Jude | 1 Thess. | 1 Thessalonians |
| Judg. | Judges | 2 Thess. | 2 Thessalonians |
| 1 Kings | 1 Kings | 1 Tim. | 1 Timothy |
| 2 Kings | 2 Kings | 2 Tim. | 2 Timothy |
| Lam. | Lamentations | Tit. | Titus |
| Lev. | Leviticus | Zech. | Zechariah |
| Luke | Luke | Zeph. | Zephaniah |
| Mal. | Malachi | | |
| Mark | Mark | | |
| Matt. | Matthew | | Bible Translations |
| Mic. | Micah | | |
| Nah. | Nahum | ESV | English Standard |
| Neh. | Nehemiah | | Version |
| Num. | Numbers | KJV | King James Version |
| Obad. | Obadiah | NASB | New American |
| 1 Pet. | 1 Peter | | Standard Bible |
| 2 Pet. | 2 Peter | | (updated) |
| Phil. | Philippians | NLT | New Living |
| Philem. | Philemon | | Translation |
| Prov. | Proverbs | NLV | New Life Version |

# Appendix 2

# An Overview of the Bible

The Bible consists of sixty-six different books written by forty authors over a period of about 1500 years. It is divided into the Hebrew Bible (Old Testament), sacred to Jews and Christians, and the New Testament, sacred to Christians. Most of the Bible originally was written in either ancient Hebrew or Greek, and it has been translated into many modern languages. Its books range in style from poetry to prose and in content from history to love songs. Among other things, it is a collection of ancient wisdom with modern applications. The books are divided into chapters and the chapters into verses. There are 31,102 verses in the entire Bible. The following is a list of the books of the Bible in the order in which they occur, followed by the standard abbreviation for each book. Some denominations and sects add other books to their list of sacred writings.

| The Hebrew Bible | | Ruth | Ruth |
|---|---|---|---|
| | | 1 Samuel | 1 Sam. |
| Genesis | Gen | 2 Samuel | 2 Sam. |
| Exodus | Exod. | 1 Kings | 1 Kings |
| Leviticus | Lev. | 2 Kings | 2 Kings |
| Numbers | Num. | 1 Chronicles | 1 Chron. |
| Deuteronomy | Deut. | 2 Chronicles | 2 Chron. |
| Joshua | Josh. | Ezra | Ezra |
| Judges | Judg. | Nehemiah | Neh. |

| | | The New Testament | |
|---|---|---|---|
| Esther | Esth. | | |
| Job | Job | | |
| Psalms | Ps. | Matthew | Matt. |
| Proverbs | Prov. | Mark | Mark |
| Ecclesiastes | Eccles. | Luke | Luke |
| Song of | | John | John |
| Solomon | Song Sol. | Acts | Acts |
| Isaiah | Isa. | Romans | Rom. |
| Jeremiah | Jer. | 1 Corinthians | 1 Cor. |
| Lamentations | Lam. | 2 Corinthians | 2 Cor. |
| Ezekiel | Ezek. | Galatians | Gal. |
| Daniel | Dan. | Ephesians | Eph. |
| Hosea | Hos. | Philippians | Phil. |
| Joel | Joel | Colossians | Col. |
| Amos | Amos | 1 Thessalonians | 1 Thess. |
| Obadiah | Obad. | 2 Thessalonians | 2 Thess. |
| Jonah | Jon. | 1 Timothy | 1 Tim. |
| Micah | Mic. | 2 Timothy | 2 Tim. |
| Nahum | Nah. | Titus | Tit. |
| Habakkuk | Hab. | Philemon | Philem. |
| Zephaniah | Zeph. | Hebrews | Heb. |
| Haggai | Hag. | James | Jas. |
| Zechariah | Zech. | 1 Peter | 1 Pet. |
| Malachi | Mal. | 2 Peter | 2 Pet. |
| | | 1 John | 1 John |
| | | 2 John | 2 John |
| | | 3 John | 3 John |
| | | Jude | Jude |
| | | Revelation | Rev. |

# Appendix 3

# The Essentials of Agriculture

The following fifteen essentials of agriculture were identified in Chapters 5–9 based on the combination of topics mentioned in the ancient wisdom of the Bible and a modern understanding of the common practices of agriculture. These essentials must be addressed by all kinds of agriculture. Sustainability is facilitated by addressing them with a long-term view.

1. Adequate natural resources must be managed in locally appropriate ways.
2. Soil must be respected and protected.
3. Water for agriculture must be managed.
4. There must be locally adapted species that can be managed for agriculture.
5. There must be effective methods of pest control.
6. There must be appropriate methods for harvesting and storing crops.
7. Livestock can make important ecological, nutritional, economic, and cultural contributions to our agricultural systems.
8. Animal welfare is a central concern of agriculture, and all livestock should be well cared for.
9. Animal traction is a model of a sustainable means of supplying the required energy for agriculture.

179

10. Agriculture must be viewed and managed in a holistic manner.
11. There must be a means of tilling the soil so that crops can be successfully grown.
12. There must be an effective means of maintaining soil fertility.
13. Farming must offer an attractive lifestyle and a means of learning how to farm so that future generations will become farmers.
14. There must be an economic system that rewards stewardship of resources and provides a way for the disadvantaged to be fed and to recover.
15. There is a religious and an ethical component of agriculture that calls for all participants to have food so that they can celebrate life.

# Notes

## Introduction

1. George Bernard Shaw, "Getting Married," in *Bernard Shaw: Collected Plays with Their Prefaces*, ed. Dan H. Laurence, vol. 3, 658 (New York: Dodd and Mead, 1975), lines spoken by Hotchkiss.

2. David Beckmann and Arthur Simon, *Grace at the Table: Ending Hunger in God's World* (Mahwah, NJ: Paulist Press, 1999).

3. Eric Schlosser, *Fast Food Nation: The Dark Side of the All-American Meal* (New York: HarperCollins, 2002).

4. Marion Nestle, *Food Politics: How the Food Industry Influences Nutrition and Health* (Berkeley: University of California Press, 2003).

5. L. Shannon Jung, *Sharing Food: Christian Practices for Enjoyment* (Minneapolis, MN: Fortress Press, 2006).

6. Stephen H. Webb, *Good Eating (The Christian Practice of Everyday Life)* (Grand Rapids, MI: Brazos Press, 2001).

7. Arthur W. Halliday and Judy Wardell Halliday, *Thin Again: A Biblical Approach to Food, Eating, and Weight Management* (Grand Rapids, MI: Revell, 2002).

8. Doris Janzen Longacre, *More-with-Less Cookbook: 25th Anniversary Edition* (Scottdale, PA: Herald Press, 2000).

9. Mary Beth Lind and Cathleen Hockman-Wert, *Simply in Season: A World Community Cookbook* (Scottdale, PA: Herald Press, 2005).

10. Jennifer Halteman Schrock, *Just Eating? Practicing Our Faith at the Table* (Pittsburgh, PA: Presbyterian Hunger Program, Presbyterian Church [USA], 2005).

11. Michael Pollan, *The Omnivore's Dilemma—A Natural History of Four Meals* (New York: Penguin Press, 2006).

12. Liberty Hyde Bailey, *The Holy Earth* (New York: C. Scribner's Sons, 1915).

13. Alastair I. MacKay, *Farming and Gardening in the Bible* (Emmaus, PA: Rodale Press, 1950).

14. David J. Evans, Ronald J. Vos, and Keith P. Wright, eds., *Biblical Holism and Agriculture: Cultivating Our Roots* (Pasadena, CA: William Carey Library, 2003).

15. Mark E. Graham, *Sustainable Agriculture: A Christian Ethic of Gratitude* (Cleveland, OH: Pilgrim Press, 2005).

16. Calvin B. DeWitt, *Caring for Creation: Responsible Stewardship of God's Handiwork* (Grand Rapids, MI: Baker Books, 1998).

17. R. J. Berry, Anne M. Clifford, Peter Harris, Michael S. Northcott, and Don Brandt, eds., *God's Stewards: The Role of Christians in Creation Care* (Monrovia, CA: World Vision, 2002).

18. John E. Carroll, *Sustainability and Spirituality* (Albany: State University of NY Press, 2004).

19. Marjorie Hope and James Young, *Voices of Hope in the Struggle to Save the Planet* (New York: Apex Press, 2000).

20. Martin Palmer and Victoria Finlay, *Faith in Conservation: New Approaches to Religions and the Environment* (Washington, DC: The World Bank, 2003).

21. Gary Gardner, *Inspiring Progress: Religions' Contributions to Sustainable Development* (New York: W.W. Norton and Company, 2006).

# Chapter 1. It Is All about Food

1. Bjørn Lomborg, *Global Crises, Global Solutions* (New York: Cambridge University Press, 2004).

2. R. M. Welch and R. D. Graham, "Agriculture: The Real Nexus for Enhancing Bioavailable Micronutrients in Food Crops," *Journal of Trace Elements in Medicine and Biology* 18 (2004): 299.

3. FAO, *What the New Figures on Hunger Mean* (Rome: Food and Agricultural Organization, United Nations, 2002). Also available online at http://www.fao.org/english/newsroom/news/2002/9703-en.html (accessed December 19, 2006).

4. World Watch Institute, "Food," *Vital Signs 2006–2007—The Trends That Are Shaping Our Future* (New York: W.W. Norton, 2006), 300–301.

5. Welch and Graham, "Agriculture," 18:299–307.

# Chapter 2. The Foundation of Stewardship

1. A comprehensive compilation of the Judeo-Christian and Muslim scriptures about agriculture is available upon request in the following publication: Gary W. Fick, *Farming by the Book: Food, Farming, and the Environment in the Bible and the Qur'an—CSS No. T05-1* (Ithaca, NY: Department of Crop and Soil Sciences, Cornell University, 2005). Also available online at http://www.dspace.library.cornell.edu/handle/1813/2550 (accessed December 19, 2006).

2. In reference to the Creation story, the phrase "truer than fact" is borrowed from an address given by Gardner Taylor at the Cornell University Baccalaureate Service in 2003.

3. Walter Brueggemann, *The Land: Place as a Gift, Promise, and Challenge in Biblical Faith,* 2d ed. (Minneapolis, MN: Fortress Press, 2002).

4. Theodore Hiebert, *The Yahwist's Landscape* (New York: Oxford University Press, 1996).

5. Lynn White Jr., "The Historical Roots of the Ecological Crisis," *Science* 155 (1967): 1203–1207.

6. Michael Pollan, in *Second Nature: A Gardener's Education* (New York: Bantam Doubleday Dell Publishing Group, 1991), 45–64.

7. Though not the main point of the present chapter, the biblical picture of the human-soil-food association is very clear here.

## Chapter 3. Ecology in the Bible

1. Judith D. Soule and Jon K. Piper, *Farming in Nature's Image: An Ecological Approach to Agriculture* (Washington, DC: Island Press, 1992), 84–86.

2. Stephen R. Gliessman, *Agroecology: The Ecology of Sustainable Food Systems,* 2d ed. (Boca Raton, FL: CRC Press, 2007).

3. Jeffry A. McNeely and Sara J. Scheer, *Ecoagriculture: Strategies to Feed the World and Save Wild Biodiversity* (Washington, DC: Island Press, 2002).

4. Diane Rickerl and Charles Francis, *Agroecosystems Analysis* (Madison, WI: American Society of Agronomy, 2004).

5. Dana L. Jackson and Laura L. Jackson, eds. *Farm as Natural Habitat: Reconnecting Food Systems with Ecosystems* (Washington, DC: Island Press, 2002).

6. The perspective of a Jewish environmentalist is recorded by soil scientist and ecologist Daniel Hillel, *The Natural History of the Bible* (New York: Columbia University Press, 2006).

7. Craig J. Hogan, "Observing the Beginning of Time," *American Scientist* 90 (2002): 420–27.

8. Lynn White Jr., "The Historical Roots of the Ecological Crisis," *Science* 155 (1967): 1203–1207.

9. Robert De Haan, "God-Given Characteristics are Celebrated," in *Biblical Holism and Agriculture: Cultivating Our Roots,* ed. D. J. Evans, R. J. Vos, and K. P. Wright, 88–90 (Pasadena, CA: William Carey Library, 2003).

## Chapter 4. Land and Climate

1. NRCS (Natural Resource Conservation Service, U.S. Department of Agriculture), "What Is a State Soil?" *State Soils,* http://www.soils.usda.gov/gallery/state_soils/ (accessed December 19, 2006).

2. Jorge Terena, "The Land Potential," in *Gaia: An Atlas of Planet Management,* ed. Norman Myers, 22–35 (New York: Anchor Books, Doubleday, 1993).

3. Walter Brueggemann, *The Land—Place as Gift, Promise, and Challenge in Biblical Faith,* 2d ed. (Minneapolis, MN: Fortress Press, 2002).

4. By the time of the compilation of the Jewish Law of Agriculture in the Mishnah (roughly 50 to 170 CE), the Jewish rabbis maintained that the Sabbath rest of the land applied only to ancient Hebrew lands, and that its main purpose was the sanctification (maintenance of holiness) of the land and of the offerings that came from the land. That does not mean that there were not also underlying agricultural principles and benefits as well. See A. J. Avery-Peck, "History of the Mishnaic Law of Agriculture," in *The Law of Agriculture in the Mishnah and Tosefta—Translation, Commentary, Theology,* ed. Jacob Neusner, 164, 169–70, 183 (Leiden, the Netherlands: Brill, 2005).

5. Daniel Hillel, *The Natural History of the Bible* (New York: Columbia University Press, 2006), 154.

6. Wes Jackson, *Becoming Native to This Place* (Lexington: University of Kentucky Press, 1994).

7. Jared Diamond, *Guns, Germs, and Steel: The Fate of Human Societies* (New York: W.W. Norton, 1997).

8. R. J. Vos, "Social Principles for 'Good' Agriculture," in *Biblical Holism and Agriculture: Cultivating Our Roots,* ed. D. J. Evans, R. J. Vos, and K. P. Wright, 53–54 (Pasadena, CA: William Carey Library, 2003).

9. Wendell Berry, *The Way of Ignorance* (Emeryville, CA: Shoemaker and Hoard, Avalon, 2005), 39–51, 91–104.

10. Ibid., 101.

# Chapter 5. Soil and Water

1. Mark E. Graham, *Sustainable Agriculture: A Christian Ethic of Gratitude* (Cleveland, OH: Pilgrim Press, 2005), 64.

2. An attractive introduction to soil science can be found in the short book by William Dubbin, *Soils* (London: Natural History Museum, 2001).

3. Walter Brueggemann, *The Land: Place as a Gift, Promise, and Challenge in Biblical Faith,* 2d ed. (Minneapolis, MN: Fortress Press, 2002).

4. Theodore Hiebert, *The Yahwist's Landscape* (New York: Oxford University Press, 1996).

5. Ibid., 32–38.

6. One example of soil degradation not mentioned in the Bible is heavy-metal contamination from various forms of industrial pollution, including land application of many kinds of municipal solid wastes.

# Chapter 6. Crops, Seeds, and Food

1. Michael Pollan, *Second Nature: A Gardener's Education* (New York: Bantam Doubleday Dell Publishing Group, 1991), 209–38.

2. Jack R. Harlan, *Crops and Man,* 2d ed. (Madison, WI: American Society of Agronomy and Crop Science Society of America, 1992), 31–60.

3. Gilbert L. Wilson, *Buffalo Bird Woman's Garden* (St. Paul: Minnesota Historical Society Press, 1984), 16. Wilson retells the story of Maxidiwiac (Buffalo Bird Woman). The original publication was in 1917 under the title *Agriculture of the Hidatsa Indians.*

# Chapter 7. Livestock and Agriculture

1. Judith D. Soule and Jon K. Piper, *Farming in Nature's Image: An Ecological Approach to Agriculture* (Washington, DC: Island Press, 1992), 124–26.

2. Sir Albert Howard, *An Agricultural Testament* (London: Oxford University Press, 1943). This book is still available through Rodale Press.

3. J. T. Reid, "The Future Role of Ruminants in Animal Production," in *Proceedings Third International Symposium on Physiology of Digestion and Metabolism in the Ruminant,* ed. A. T. Phillipson, 1–22 (Newcastle upon Tyne, UK: Oriel Press, 1970).

4. J. W. Oltjen and J. L. Beckett, "Role of Ruminant Livestock in Sustainable Agricultural Systems," *Journal of Animal Science* 74 (1996): 1406–1409.

5. Jo Robinson, *Why Grassfed Is Best! The Surprising Benefits of Grassfed Meat, Eggs, and Dairy Products* (Vashon, WA: Vashon Island Press, 2000). A long list of relevant references can be found at http://www.eatwild.com/references.html (accessed December 19, 2006).

6. Gary W. Fick and E. A. Clark, "The Future of Grass for Dairy Cattle," in *Grass for Dairy Cattle,* ed. J. H. Cherney and D. J. R. Cherney, 5 (Wallingford, UK: CABI Publishing, 1998).

7. Ibid., 4.

8. Daniel E. Vasey, *An Ecological History of Agriculture:10,000 b.c.–a.d. 10,000* (Ames: Iowa State University Press, 1992), 150–160.

9. Jared Diamond, *Guns, Germs, and Steel: The Fate of Human Societies* (New York: W.W. Norton, 1997), 25–32.

10. Peter R. Cheeke, *Impacts of Livestock Production on Society: Diet/ Health and the Environment* (Danville, IL: Interstate Publishers, 1993).

11. Some translations (NASB, NLV) prohibit eating meat from an animal "torn to pieces in the field" instead of "torn by beasts in the field."

12. The first few days of the nursing period guarded the health of the young animal and that of its mother. Modern physiological study has shown that colostrums in the first milk fortify the immune system of the newborn, while the process of nursing stimulates the contraction of the uterus and the mother's recovery from the stress of birth.

13. Theodore Hiebert, *The Yahwist's Landscape* (New York: Oxford University Press, 1996), 60.

14. It is possible that the ancient Hebrews managed the danger from bulls simply by using most of them at an early age for offerings and food.

15. Mark E. Graham, *Sustainable Agriculture: A Christian Ethic of Gratitude* (Cleveland, OH: Pilgrim Press, 2005), 188–89.

16. Robert De Haan, "God-Given Characteristics Are Celebrated," in *Biblical Holism and Agriculture: Cultivating Our Roots*, ed. D. J. Evans, R. J. Vos, and K. P. Wright, 88–89 (Pasadena, CA: William Carey Library, 2003).

17. Michael Pollan, *The Omnivore's Dilemma—A Natural History of Four Meals* (New York: Penguin Press, 2006), 319–25.

# Chapter 8. Farming Systems and the Practices of Agriculture

1. Brian Baker, "Brief History of Organic Farming and the National Organic Program," *Organic Farming Compliance Handbook*, http://www.sarep.ucdavis.edu/organic/complianceguide/ (accessed December 19, 2006).

2. Yao-Chi Lu, J. R. Teasdale, and Wen-Yuen Huang, "An Economic and Environmental Tradeoff Analysis of Sustainable Agriculture Cropping Systems," *Journal of Sustainable Agriculture* 22:3 (2003): 25–41.

3. M. M. Gregory, K. L. Shea, and E. B. Bakko, "Comparing Agroecosystems: Effects of Cropping and Tillage Patterns on Soil, Water, Energy Use, and Productivity," *Renewable Agriculture and Food Systems* 20 (2005): 81–90.

4. Paul Kristiansen, Acram Taji, and John Reganold, eds., *Organic Agriculture—A Global Perspective* (Ithaca, NY: Cornell University Press, 2006).

5. J. H. Cherney, D. J. R. Cherney, and T. W. Bruulsema., "Potassium Management," in *Grass for Dairy Cattle*, ed. J. H. Cherney and D. J. R. Cherney, 149–53 (Wallingford, UK: CABI Publishing, 1998).

6. Richard L. Thompson and Sharon N. Thompson, *Alternatives in Agriculture: 2003 Report* (Boone, IA: Thompson on-Farm Research, 2004), 1–2.

7. Wendell Berry, *Home Economics* (San Francisco: North Point Press, 1987), 56–57.

8. Alastair I. MacKay, *Farming and Gardening in the Bible* (Emmaus, PA: Rodale Press, 1950), 128–29.

# Chapter 9. The Culture of Agriculture

1. Depending on the location, squash or pumpkin may be grown as the third crop of the Three Sisters. Squash and pumpkin are close botanical and nutritional relatives.

2. Xenoergonics is a synthesized word with Greek roots, meaning "foreign working." It refers to the immigrant and often migrant labor force exploited by our modern food and agricultural system.

3. Theodore Hiebert, *The Yahwist's Landscape* (New York: Oxford University Press, 1996), 32–35.

4. It also is possible to see the foundation of art in the biblical story that begins and ends in a garden (Gen. 2:8; Rev. 22:1–2). The art of gardening is well illustrated by Michael Pollan, *Second Nature: A Gardener's Education* (New York: Bantam Doubleday Dell Publishing Group, 1991). See the chapters "Into the Rose Garden," 93–115, and "Planting a Tree," 178–208.

5. D. L. Miller, "Agriculture and the Kingdom of God," in *Biblical Holism and Agriculture: Cultivating Our Roots,* ed. D. J. Evans, R. J. Vos, and K. P. Wright, 152 (Pasadena, CA: William Carey Library, 2003).

6. David B. Grigg, *The Agricultural Systems of the World: An Evolutionary Approach* (London: Cambridge University Press, 1974), 187–189.

7. Kara Unger Ball, "Is Our Agricultural House Built on Sand? Biblical Holism in Agriculture and the Assumption of Monotonicity in the Utility Function," in *Biblical Holism and Agriculture: Cultivating Our Roots,* ed. D. J. Evans, R. J. Vos, and K. P. Wright, 251–65 (Pasadena, CA: William Carey Library, 2003).

8. The Jubilee cycle is variously regarded to have been forty-nine or fifty years by modern scholars. See, for example, J. S. Bergsma, "Once Again, the Jubilee, Every 49 or 50 Years?" *Vertus Testamentum* 55 (2005): 121–25.

9. Miller, "Agriculture," 147.

## Chapter 10. Abuse, Poverty, and Women

1. Ikerd develops the same argument in terms of exploitation. See John Ikerd, *Sustainable Capitalism: A Matter of Common Sense* (Bloomfield, CT: Kumarian Press, 2005), 160.

2. Lynn R. Brown et al., "Women as Producers, Gatekeepers, and Shock Absorbers," in *The Unfinished Agenda: Perspectives on Overcoming Hunger, Poverty, and Environmental Degradation,* ed. Per Pinstrup-Andersen and Rajul Pandya-Lorch, 206 (Washington, DC: International Food Policy Research Institute, 2001).

3. David Beckmann and Arthur Simon, *Grace at the Table: Ending Hunger in God's World* (Mahwah, NJ: Paulist Press, 1999), 53–57.

4. Jerome Binde, "Ready for the 21st Century?" *Futures Bulletin* 25:1 (1999): 1. Also available online at http://www.wfsf.org/pub/publications/bulletin/mar_apr00.pdf (accessed December 19, 2006).

5. P. Webb and K. Weinberger, eds., *Women Farmers: Enhancing Rights, Recognition, and Productivity* (Frankfurt, Germany: Lang Verlag, 2001).

6. Janet Henshall Momse, "Women Farmers: Environmental Managers of the World," in *Population, Land Management, and Environmental Change,* ed. J. I. Uitto and Akiko Ono, 28–33 (Tokyo: United Nations

University, 1996), available online at http://www.unu.edu/unupress/unupbooks/uu03pe/uu03pe00.htm (accessed December 19, 2006).

7. IFAP-FIPA, *International Conference on Women in Agriculture, 19–21 November 2003, Manilla, Philippines* (Paris: IFAP-FIPA, 2003), http://www.ifap.org/issues/ 2confwom03/report.html (accessed December 19, 2006).

8. Ronald J. Sider, *Rich Christians in an Age of Hunger* (Nashville, TN: Word Publishing, 1990).

9. Craig L. Blomberg, *Neither Poverty nor Riches: A Biblical Theology of Material Possessions* (Grand Rapids, MI: W. B. Eerdmans, 1999).

10. Ben Witherington III, *Women in the Earliest Churches* (Cambridge: Cambridge University Press, 1988).

11. Sarah Sumner, *Men and Women in the Church: Building Consensus of Christian Leadership* (Downers Grove, IL: Inter-Varsity Press, 2003).

12. Mark E. Graham, *Sustainable Agriculture: A Christian Ethic of Gratitude* (Cleveland, OH: Pilgrim Press, 2005), 39.

13. Eric Schlosser, *Fast Food Nation: The Dark Side of the All-American Meal* (New York: HarperCollins, 2002).

14. Ikerd, *Sustainable Capitalism: A Matter of Common Sense*, 2.

15. Ibid., 139–76.

16. Ibid., 3.

17. Kara Unger Ball, "Is Our Agricultural House Built on Sand? Biblical Holism in Agriculture and the Assumption of Monotonicity in the Utility Function," in *Biblical Holism and Agriculture: Cultivating Our Roots*, ed. D. J. Evans, R. J. Vos, and K. P. Wright, 259–61 (Pasadena, CA: William Carey Library, 2003).

18. Ikerd, *Sustainable Capitalism: A Matter of Common Sense*, 199.

19. Michael Oye, "The Bible as Ethical Standard for Appraising Modern Agricultural Practices," in *Biblical Holism and Agriculture: Cultivating Our Roots*, ed. D. J. Evans, R. J. Vos, and K. P. Wright, 187–202 (Pasadena, CA: William Carey Library, 2003).

20. Wendell Berry, *"Recollected Essays, 1965–1980,"* (San Francisco: North Point Press, 1981), 215.

21. Walter Brueggemann, *The Land: Place as a Gift, Promise, and Challenge in Biblical Faith*, 2d ed. (Minneapolis, MN: Fortress Press, 2002), 175.

22. Heifer Project International, http://www.heifer.org/ (accessed December 19, 2006).

23. ICRA (International Catholic Rural Association), http://www.icra-agrimissio.org/ (accessed December 19, 2006).

24. Lutheran World Relief, http://www.lwr.org/ourwork/development/index.asp (accessed December 19, 2006).

25. Mennonite Central Committee, http://www.mcc.org/ (accessed December 19, 2006).

26. Sameena Nazir, "Challenging Inequality—Obstacles and Opportunities towards Women's Rights in the Middle East and North Africa," in *Women's Rights in the Middle East and North Africa*, ed. Sameena Nazir and Leigh Tomppert, 1–14 (New York: Freedom House, 2005).

# Chapter 11. Food, Starvation, Obesity, and Diet

1. FAO, *What the New Figures on Hunger Mean* (Rome: Food and Agricultural Organization, United Nations, 2002). Also available online at http://www.fao.org/english/newsroom/news/2002/9703-en.html (accessed December 19, 2006).

2. David Beckmann and Arthur Simon, *Grace at the Table: Ending Hunger in God's World* (Mahwah, NJ: Paulist Press, 1999), 201–206.

3. John L. Peters, *Cry Dignity* (Oklahoma City, OK: World Neighbors, 1979).

4. R. L. Wixom, "Synthesis of Religious Faiths and Sustainable Development," in *Environmental Challenges for Higher Education: Integrating Sustainability into Academic Programs*, ed. R. L. Wixom, L. L. Gould, Susan Schmidt, and Louis Cox, 197–215 (Burlington, VT: Friends Committee on Unity with Nature, 1996).

5. L. Shannon Jung, *Sharing Food: Christian Practices for Enjoyment* (Minneapolis, MN: Fortress Press, 2006).

6. Marion Nestle, *Food Politics: How the Food Industry Influences Nutrition and Health* (Berkeley: University of California Press, 2003), 7–8.

7. Ibid., 21–23.

8. Michael Pollan, *The Omnivore's Dilemma—A Natural History of Four Meals* (New York: Penguin Press, 2006), 100–108.

9. Jo Robinson, *Why Grassfed Is Best! The Surprising Benefits of Grassfed Meat, Eggs, and Dairy Products* (Vashon, WA: Vashon Island Press, 2000), 11–28, 43–48.

10. United States Department of Agriculture, *MyPyramid.gov—Steps to a Healthier You*, http://www.mypyramid.gov/ (accessed December 19, 2006).

11. Nestle, *Food Politics*, 84–91.

12. Frances M. Lappe, *Diet for a Small Planet* (New York: Ballantine Books, 1991).

13. Doris Janzen Longacre, *More-with-Less Cookbook: 25th Anniversary Edition* (Scottdale, PA: Herald Press, 2000).

14. Mark E. Graham, *Sustainable Agriculture: A Christian Ethic of Gratitude* (Cleveland, OH: Pilgrim Press, 2005).

# Chapter 12. Agricultural Sustainability

1. C. E. Beus and R. E. Dunlap, "Conventional versus Alternative Agriculture: The Paradigmatic Roots of the Debate," *Rural Sociology* 55 (1990): 590–616.

2. Kerry H. Whiteside, *Precautionary Politics* (Cambridge, MA: MIT Press, 2006).

3. M. B. Chiappe and C. B. Flora, "Gendered Elements of the Alternative Agriculture Paradigm," *Rural Sociology* 63 (1998): 372–93.

4. Beus and Dunlap, "Conventional versus Alternative Agriculture," 590–616.

5. R. W. Elmore, "Our Relationship with the Ecosystem and Its Impact on Sustainable Agriculture," *Journal of Production Agriculture* 9 (1996): 3–4, 42–45.

6. David J. Evans, Ronald J. Vos, and Keith P. Wright, eds., *Biblical Holism and Agriculture: Cultivating Our Roots* (Pasadena, CA: William Carey Library, 2003).

7. Martin Palmer and Victoria Finlay, *Faith in Conservation: New Approaches to Religions and the Environment* (Washington, DC: The World Bank, 2003).

8. Gary Gardner, *Inspiring Progress: Religions' Contributions to Sustainable Development* (New York: W.W. Norton and Company, 2006).

9. T. E. Crews, C. L. Mohler, and A. G. Power, "Energetics and Ecosystem Integrity: The Defining Principles of Sustainable Agriculture," *American Journal of Alternative Agriculture* 6 (1991): 146–49.

10. David W. Orr, *Ecological Literacy: Education and the Transition to a Postmodern World* (Albany: State University of New York Press, 1992), 97–108.

11. Jan Wojcik, *The Arguments of Agriculture: A Casebook in Contemporary Agricultural Controversy* (Lafayette, IN: Purdue University Press, 1989).

12. Ikerd also has suggested a similar alignment of topics. See John Ikerd, *Sustainable Capitalism: A Matter of Common Sense* (Bloomfield CT: Kumarian Press, 2005), 85.

13. A whole book on the subject is by Jared Diamond, *Guns, Germs, and Steel: The Fate of Human Societies* (New York: W.W. Norton, 1997).

14. Gilbert W. Gillespie Jr., "Sustainability: Too Many Legs Spoil the Stool," *SANET-MG@LISTS.IFAS.UFL.EDU*, sent May 1, 2002.

15. Wes Jackson, *Becoming Native to This Place* (Lexington: University of Kentucky Press, 1994), 87–103.

16. Wendell Berry, *The Way of Ignorance* (Emeryville, CA: Shoemaker and Hoard, Avalon, 2005), 45.

17. Barry Commoner, *The Closing Circle: Nature, Man, and Technology* (New York: Alfred A. Knopf, 1971), 33.

18. Practical eating that supports agricultural sustainability and social justice is described in several publications mentioned in the introduction, including the recent study guide by Jennifer Halteman Schrock, *Just Eating? Practicing Our Faith at the Table* (Pittsburgh, PA: Presbyterian Hunger Program, Presbyterian Church [USA], 2005).

19. David Andrews, "Eating as a Moral Act," *Eating as a Moral Act: Ethics and Power from Agrarianism to Consumerism Symposium and Saul O. Sidore Memorial Lecture Series Proceedings, April 25–27, 2004* (Durham: Office of Sustainability Programs, University of New Hampshire, 2004), 24.

# Bibliography

Andrews, David. "Eating as a Moral Act." *Eating as a Moral Act: Ethics and Power from Agrarianism to Consumerism Symposium and Saul O. Sidore Memorial Lecture Series Proceedings, April 25–27, 2004*. Durham: Office of Sustainability Programs, University of New Hampshire, 2004.

Avery-Peck, A. J. "History of the Mishnaic Law of Agriculture." In *The Law of Agriculture in the Mishnah and the Tosefta—Translation, Commentary, Theology*, edited by Jacob Neusner, 341–97. Leiden, the Netherlands: Brill, 2005.

Bailey, Liberty Hyde. *The Holy Earth*. New York: C. Scribner's Sons, 1915.

Baker, Brian. "Brief History of Organic Farming and the National Organic Program." *Organic Farming Compliance Handbook*. http://www.sarep. ucdavis.edu/organic/complianceguide/ (accessed December 19, 2006).

Beckmann, David, and Arthur Simon. *Grace at the Table: Ending Hunger in God's World*. Mahwah, NJ: Paulist Press, 1999.

Bergsma, J. S. "Once Again, the Jubilee, Every 49 or 50 Years?" *Vertus Testamentum* 55 (2005): 121–25.

Berry, R. J., ed. *The Care of Creation: Focusing Concern and Action*. Downers Grove, IL: Inter-Varsity Press, 2000.

Berry, R. J., Anne M. Clifford, Peter Harris, Michael S. Northcott, and Don Brandt, eds. *God's Stewards: The Role of Christians in Creation Care*. Monrovia, CA: World Vision, 2002.

Berry, Wendell. *Recollected Essays, 1965–1980*. San Francisco: North Point Press, 1981.

Berry, Wendell. *Home Economics*. San Francisco: North Point Press, 1987.

Berry, Wendell. *The Way of Ignorance*. Emeryville, CA: Shoemaker and Hoard, Avalon, 2005.

Beus, C. E., and R. E. Dunlap. "Conventional versus Alternative Agriculture: The Paradigmatic Roots of the Debate." *Rural Sociology* 55 (1990): 590–616.

Binde, Jerome. "Ready for the 21st Century?" *Futures Bulletin* 25:1 (1999): 1–20. Also available online at http://www.wfsf.org/pub/publications/bulletin/mar_apr00.pdf (accessed December 19, 2006).

Blomberg, Craig L. *Neither Poverty nor Riches: A Biblical Theology of Material Possessions*. Grand Rapids, MI: W. B. Eerdmans, 1999.

Bouma-Prediger, Steven. *For the Beauty of the Earth: A Christian Vision for Creation Care*. Grand Rapids, MI: Baker Academic, 2001.

Brown, Lynn R., Hilary Feldstein, Lawrence Haddad, Christine Peña, and Agnes Quisumbing. "Women as Producers, Gatekeepers, and Shock Absorbers." In *The Unfinished Agenda: Perspectives on Overcoming Hunger, Poverty, and Environmental Degradation*, edited by Per Pinstrup-Andersen and Rajul Pandya-Lorch, 205–209. Washington, DC: International Food Policy Research Institute, 2001.

Brueggemann, Walter. *The Land: Place as a Gift, Promise, and Challenge in Biblical Faith*. 2d ed. Minneapolis, MN: Fortress Press, 2002.

Carroll, John E. *Sustainability and Spirituality*. Albany: State University of New York Press, 2004.

Carroll, John E. *The Wisdom of Small Farms and Local Food: Aldo Leopold's Land Ethic and Sustainable Agriculture*. Durham: University of New Hampshire, 2006.

Cheeke, Peter R. *Impacts of Livestock Production on Society: Diet/Health and the Environment*. Danville, IL: Interstate Publishers, 1993.

Cherney, J. H., D. J. R. Cherney, and T. W. Bruulsema. "Potassium Management." In *Grass for Dairy Cattle*, edited by J. H. Cherney and D. J. R. Cherney, 137–60. Wallingford, UK: CABI Publishing, 1998.

Chiappe, M. B., and C. B. Flora. "Gendered Elements of the Alternative Agriculture Paradigm." *Rural Sociology* 63 (1998): 372–93.

Commoner, Barry. *The Closing Circle: Nature, Man, and Technology*. New York: Alfred A. Knopf, 1971.

Crews, T. E., C. L. Mohler, and A.G. Power. "Energetics and Ecosystem Integrity: The Defining Principles of Sustainable Agriculture." *American Journal of Alternative Agriculture* 6 (1991): 146–49.

De Haan, Robert. "God-Given Characteristics are Celebrated." In *Biblical Holism and Agriculture: Cultivating Our Roots*, edited by D. J. Evans, R. J. Vos, and K. P. Wright, 81–97. Pasadena, CA: William Carey Library, 2003.

DeWitt, Calvin B. *Caring for Creation: Responsible Stewardship of God's Handiwork*. Grand Rapids, MI: Baker Books, 1998.

DeWitt, Calvin B. *Earth-Wise: A Biblical Response to Environmental Issues*. Grand Rapids, MI: Faith Alive Christian Resources, 1994.

DeWitt, Calvin B., ed. *The Environment and the Christian: What Does the New Testament Say about the Environment?* Grand Rapids, MI: Baker Books, 1991.

Diamond, Jared. *Guns, Germs, and Steel: The Fate of Human Societies*. New York: W.W. Norton, 1997.

Dubbin, William. *Soils*. London: Natural History Museum, 2001.

Elmore, R. W. "Our Relationship with the Ecosystem and Its Impact on Sustainable Agriculture." *Journal of Production Agriculture* 9 (1996): 3–4 , 42–45.

Evans, David J., Ronald J. Vos, and Keith P. Wright, eds. *Biblical Holism and Agriculture: Cultivating Our Roots.* Pasadena, CA: William Carey Library, 2003.

FAO. *What the New Figures on Hunger Mean.* Rome: Food and Agricultural Organization, United Nations, 2002. Also available online at http://www.fao. org/english/newsroom/news/2002/9703-en.html (accessed December 19, 2006).

Fick, Gary W. *Farming by the Book: Food, Farming, and the Environment in the Bible and the Qur'an—CSS No. T05-1.* Ithaca, NY: Department of Crop and Soil Sciences, Cornell University, 2005. Also available online at http://www.dspace. library.cornell.edu/handle/1813/2550 (accessed December 19, 2006).

Fick, Gary W., and E. A. Clark. "The Future of Grass for Dairy Cattle." In *Grass for Dairy Cattle,* edited by J. H. Cherney and D. J. R. Cherney, 1–22. Wallingford, UK: CABI Publishing, 1998.

Gardner, Gary. *Inspiring Progress: Religions' Contributions to Sustainable Development.* New York: W. W. Norton and Company, 2006.

Gillespie, Gilbert W., Jr. "Sustainability: Too Many Legs Spoil the Stool." *SANET-MG@LISTS.IFAS. UFL.EDU,* sent May 1, 2002.

Gliessman, Stephen R. *Agroecology: The Ecology of Sustainable Food Systems.* 2d ed. Boca Raton, FL: CRC Press, 2007.

Gold, M.V. "Sustainable Agriculture: Definitions and Terms." *USDA Special Reference Brief 94-05.* Beltsville, MD: National Agriculture Library, 1994.

Graham, Mark E. *Sustainable Agriculture: A Christian Ethic of Gratitude.* Cleveland, OH: Pilgrim Press, 2005.

Gregory, M. M., K. L. Shea, and E. B. Bakko. "Comparing Agroecosystems: Effects of Cropping and Tillage Patterns on Soil, Water, Energy Use, and Productivity." *Renewable Agriculture and Food Systems* 20 (2005): 81–90.

Grigg, David B. *The Agricultural Systems of the World: An Evolutionary Approach.* London: Cambridge University Press, 1974.

Halliday, Arthur W., and Judy Wardell Halliday. *Thin Again: A Biblical Approach to Food, Eating, and Weight Management.* Grand Rapids, MI: Revell, 2002.

Harlan, Jack R. *Crops and Man,* 2d ed. Madison, WI: American Society of Agronomy and Crop Science Society of America, 1992.

Heifer Project International. http://www.heifer.org/ (accessed December 19, 2006).

Hiebert, Theodore. *The Yahwist's Landscape.* New York: Oxford University Press, 1996.

Hillel, Daniel. *The Natural History of the Bible.* New York: Columbia University Press, 2006.

Hogan, Craig J. "Observing the Beginning of Time." *American Scientist* 90 (2002): 420–27.

Hope, Marjorie, and James Young. *Voices of Hope in the Struggle to Save the Planet.* New York: Apex Press, 2000.

Howard, Sir Albert. *An Agricultural Testament.* London: Oxford University Press, 1943.

ICRA (International Catholic Rural Association). http://www.icra-agrimissio.org/ (accessed December 19, 2006).

IFAP-FIPA. *International Conference on Women in Agriculture 19–21 November 2003, Manilla, Philippines.* Paris: IFAP-FIPA, 2003. http://www.ifap.org/issues/2confwom03/report.html (accessed December 19, 2006).

Ikerd, John. *Sustainable Capitalism: A Matter of Common Sense.* Bloomfield, CT: Kumarian Press, 2005.

Jackson, Dana L., and Laura L. Jackson, eds. *Farm as Natural Habitat: Reconnecting Food Systems with Ecosystems.* Washington, DC: Island Press, 2002.

Jackson, Wes. *Becoming Native to This Place.* Lexington: University of Kentucky Press, 1994.

Jung, L. Shannon. *Sharing Food: Christian Practices for Enjoyment.* Minneapolis, MN: Fortress Press, 2006.

Kristiansen, Paul, Acram Taji, and John Reganold, eds. *Organic Agriculture—A Global Perspective.* Ithaca, NY: Cornell University Press, 2006.

Lappe, Frances M. *Diet for a Small Planet.* New York: Ballantine Books, 1991.

Lind, Mary Beth, and Cathleen Hockman-Wert. *Simply in Season: A World Community Cookbook.* Scottdale, PA: Herald Press, 2005.

Lomborg, Bjørn. *Global Crises, Global Solutions.* New York: Cambridge University Press, 2004.

Longacre, Doris Janzen. *More-with-Less Cookbook: 25th Anniversary Edition.* Scottdale, PA: Herald Press, 2000.

Lu, Yao-Chi, J. R. Teasdale, and Wen-Yuen Huang. "An Economic and Environmental Tradeoff Analysis of Sustainable Agriculture Cropping Systems." *Journal of Sustainable Agriculture* 22:3 (2003): 25–41.

Lutheran World Relief. http://www.lwr.org/ourwork/ development/index.asp (accessed December 19, 2006).

MacKay, Alastair I. *Farming and Gardening in the Bible.* Emmaus, PA: Rodale Press, 1950.

McNeely, Jeffry A., and Sara J. Scheer. *Ecoagriculture: Strategies to Feed the World and Save Wild Biodiversity.* Washington, DC: Island Press, 2002.

Mennonite Central Committee. http://www.mcc.org/ (accessed December 19, 2006).

Miller, D. L. "Agriculture and the Kingdom of God." In *Biblical Holism and Agriculture: Cultivating Our Roots,* edited by D. J. Evans, R. J. Vos, and K. P. Wright, 141–72. Pasadena, CA: William Carey Library, 2003.

Momse, Janet Henshall. "Women Farmers: Environmental Managers of the World." In *Population, Land Management, and Environmental Change,* edited by J. I. Uitto and Akiko Ono, 28–33. Tokyo: United Nations University, 1996. Also available online at http://www.unu.edu/unupress/unupbooks/uu03pe/uu03pe00.htm (accessed December 19, 2006).

Nazir, Sameena. "Challenging Inequality—Obstacles and Opportunities towards Women's Rights in the Middle East and North Africa." In *Women's Rights in the Middle East and North Africa,* edited by Sameena Nazir and Leigh Tomppert, 1–14. New York: Freedom House, 2005.

Nestle, Marion. *Food Politics: How the Food Industry Influences Nutrition and Health.* Berkeley: University of California Press, 2003.

NRCS (Natural Resource Conservation Service, U.S. Department of Agriculture). "What Is a State Soil?" *State Soils.* http://www.soils.usda.gov/gallery/state_soils/ (accessed December 19, 2006).

Oltjen, J. W., and J. L. Beckett. "Role of Ruminant Livestock in Sustainable Agricultural Systems." *Journal of Animal Science* 74 (1996): 1406–1409.

Orr, David W. *Ecological Literacy: Education and the Transition to a Postmodern World.* Albany: State University of New York Press, 1992.

Oye, Michael. "The Bible as Ethical Standard for Appraising Modern Agricultural Practices." In *Biblical Holism and Agriculture: Cultivating Our Roots,* edited by D. J. Evans, R. J. Vos, and K. P. Wright, 187–202. Pasadena, CA: William Carey Library, 2003.

Palmer, Martin, and Victoria Finlay. *Faith in Conservation: New Approaches to Religions and the Environment.* Washington, DC: The World Bank, 2003.

Peters, John L. *Cry Dignity.* Oklahoma City, OK: World Neighbors, 1979.

Pollan, Michael. *The Omnivore's Dilemma—A Natural History of Four Meals.* New York: Penguin Press, 2006.

Pollan, Michael. *Second Nature: A Gardener's Education.* New York: Bantam Doubleday Dell Publishing Group, 1991.

Reid, J. T. "The Future Role of Ruminants in Animal Production." In *Proceedings Third International Symposium on Physiology of Digestion and Metabolism in the Ruminant,* edited by A. T. Phillipson, 1–22. Newcastle upon Tyne, UK: Oriel Press, 1970.

Rickerl, Diane, and Charles Francis. *Agroecosystems Analysis.* Madison, WI: American Society of Agronomy, 2004.

Robinson, Jo. *Why Grassfed Is Best! The Surprising Benefits of Grassfed Meat, Eggs, and Dairy Products.* Vashon, WA: Vashon Island Press, 2000.

Schlosser, Eric. *Fast Food Nation: The Dark Side of the All-American Meal.* New York: HarperCollins, 2002.

Schrock, Jennifer Halteman. *Just Eating? Practicing Our Faith at the Table.* Pittsburgh, PA: Presbyterian Hunger Program, Presbyterian Church (USA), 2005.

Shaw, George Bernard. "Getting Married." In *Bernard Shaw: Collected Plays with Their Prefaces,* edited by Dan H. Laurence, vol. 3, 449–668. New York: Dodd and Mead, 1975.

Sider, Ronald J. *Rich Christians in an Age of Hunger.* Nashville, TN: Word Publishing, 1990.

Soule, Judith D., and Jon K. Piper. *Farming in Nature's Image: An Ecological Approach to Agriculture.* Washington, DC: Island Press, 1992.

Sumner, Sarah. *Men and Women in the Church: Building Consensus of Christian Leadership.* Downers Grove, IL: Inter-Varsity Press, 2003.

Terena, Jorge. "The Land Potential." In *Gaia: An Atlas of Planet Management,* ed. Norman Myers, 22–35. New York: Anchor Books, Doubleday, 1993.

Thompson, Richard L., and Sharon N. Thompson. *Alternatives in Agriculture: 2003 Report.* Boone, IA: Thompson on-Farm Research, 2004.

Unger Ball, Kara. "Is Our Agricultural House Built on Sand? Biblical Holism in Agriculture and the Assumption of Monotonicity in the Utility Function." In *Biblical Holism and Agriculture: Cultivating Our Roots,* edited by D. J. Evans, R. J. Vos, and K. P. Wright, 251–65. Pasadena, CA: William Carey Library, 2003.

United States Conference of Catholic Bishops. *For I Was Hungry and You Gave Me Food: Catholic Reflections on Food, Farmers, and Farmworkers.* Washington, DC: USCCB, 2003.

United States Department of Agriculture. *MyPyramid.gov—Steps to a Healthier You.* http://www.mypyramid.gov/ (accessed December 19, 2006).

Van Dyke, Fred, David C. Mahan, Joseph K. Sheldon, and Raymond H. Brand. *Redeeming Creation: The Biblical Basis for Environmental Stewardship.* Downers Grove, IL: Inter-Varsity Press, 1996.

Vasey, Daniel E. *An Ecological History of Agriculture: 10,000 b.c.–a.d. 10,000.* Ames: Iowa State University Press, 1992.

Vos, R. J. "Social Principles for 'Good' Agriculture." In *Biblical Holism and Agriculture: Cultivating Our Roots,* edited by D. J. Evans, R. J. Vos, and K. P. Wright, 43–63. Pasadena, CA: William Carey Library, 2003.

Webb, P., and K. Weinberger, eds. *Women Farmers: Enhancing Rights, Recognition, and Productivity.* Frankfurt, Germany: Lang Verlag, 2001.

Webb, Stephen H. *Good Eating (The Christian Practice of Everyday Life).* Grand Rapids, MI: Brazos Press, 2001.

Welch, R. M., and R. D. Graham. "Agriculture: The Real Nexus for Enhancing Bioavailable Micronutrients in Food Crops." *Journal of Trace Elements in Medicine and Biology* 18 (2004): 299–307.

White, Lynn, Jr. "The Historical Roots of the Ecological Crisis." *Science* 155 (1967): 1203–1207.

Whiteside, Kerry H. *Precautionary Politics.* Cambridge, MA: MIT Press, 2006.

Wilson, Gilbert L. *Buffalo Bird Woman's Garden.* St. Paul: Minnesota Historical Society Press, 1984.

Witherington, Ben, III. *Women in the Earliest Churches.* Cambridge: Cambridge University Press, 1988.

Wixom, R. L. "Synthesis of Religious Faiths and Sustainable Development." In *Environmental Challenges for Higher Education: Integrating Sustainability into Academic Programs,* edited by R. L. Wixom, L. L. Gould, Susan Schmidt, and Louis Cox, 197–215. Burlington, VT: Friends Committee on Unity with Nature, 1996.

Wojcik, Jan. *The Arguments of Agriculture: A Casebook in Contemporary Agricultural Controversy.* Lafayette, IN: Purdue University Press, 1989.

World Watch Institute. "Food." *Vital Signs 2006–2007—The Trends That Are Shaping Our Future.* New York: W. W. Norton, 2006.

Shelton State Libraries
Shelton State Community College

# Index

# Index of Bible References

Bolded page numbers have quotations